Engineering Documentation Control/ Configuration Management Standards Manual

Scrivener Publishing
100 Cummings Center, Suite 541J
Beverly, MA 01915-6106
Publishers at Scrivener
Martin Scrivener (martin@scrivenerpublishing.com)
Phillip Carmical (pcarmical@scrivenerpublishing.com)

Engineering Documentation Control/Configuration Management Standards Manual

Policies, Procedures, Flow Diagrams, Forms and Form Instructions for Product Manufacturing Companies

Frank B. Watts

Scrivener Publishing

WILEY

This edition first published 2018 by John Wiley & Sons, Inc., 111 River Street, Hoboken, NJ 07030, USA and Scrivener Publishing LLC, 100 Cummings Center, Suite 541J, Beverly, MA 01915, USA

For more information about Scrivener publications please visit www.scrivenerpublishing.com.

Wiley Global Headquarters
111 River Street, Hoboken, NJ 07030, USA

For details of our global editorial offices, customer services, and more information about Wiley products visit us at www.wiley.com.

Library of Congress Cataloging-in-Publication Data
ISBN 978-1-119-47903-1

Cover images: Pixabay.Com
Cover design by: Russell Richardson

Set in size of 12pt and Minion Pro by Exeter Premedia Services Private Ltd., Chennai, India

10 9 8 7 6 5 4 3 2 1

Contents

Introduction

Engineering Documentation Control — sometimes called Configuration Management or Product Lifecycle Management — is a key ingredient to world-class product manufacturing.

Most product manufacturing companies suffer from the "wall syndrome." The "manufacturing side" bought ERP/Supply Chain tools; the "engineering side" bought CAD/PDM/PLM. Those software systems do not generally "talk" to each other. The engineering folks are, by and large, analytical and cautious (Ready . . . Aim . . . Fire); the manufacturing folks are, by and large, shakers, movers, and doers (Fire-Aim, Fire-Aim, Fire-Aim). The people often do not communicate very well. The manufacturing folks say that engineering "throws it over the wall." The development engineering folks say that you cannot find anyone who knows how the product will be processed. This situation often results in a huge "wall" or "gap" between engineering and operations. A significant part of bridging the gap is to develop "make sense" standards.

There is a very scary tendency in industry today; after the loose identification of a problem, the tendency is to seek a software application solution. App mania!

Software programs may help after you understand the job that needs to be done and what process flow is best for you. Something more substantial (than software) is needed between engineering and operations — Configuration Management. This author may use various terms but will most often use "CM," as that is the term which is becoming most used in industry.

In order to achieve best-in-class CM it is necessary to document what you do, do what you document and, preferably, continuously improve what you do. Improve what you do by continuous improvement or by reengineering the processes. Of course, brand new startups can leap directly into best in class with these standards.

Yes, there is a gap between engineering and the rest of the company. With this manual you can bridge that gap. It also requires a dedicated and driven CM manager and/or engineering services director.

There are many commercial, military and agency standards — enough to boggle the mind. They will tell you what the expectations are for the "outcome" they desire — but give little help in the "how to." This manual is directed at how to achieve best-in-class processes which are "make sense," fast, accurate, efficient, effective, measured and well understood.

With this manual, you have the best of the best management practices for the configuration management processes. They also go a long way toward satisfying Total Quality Management, FDA, GMP, Lean CM and ISO/QS/AS 9XXX process documentation requirements. The one requirement common to all those standards is to document the processes and to do what you document.

This manual has been under development and improvement for many years. It was sold privately by the author for several years. Many copies have been used by engineering documentation control (EDC)/configuration management (CM) managers by editing to suit their particular product manufacturing environment.

This Standards Manual should be an invaluable guide in developing your standards. It will save you many man-hours of research, development, writing, form design and procedural flow design time.

Further understanding of these practices can be obtained by reading the Engineering Documentation Control Handbook and/or CM for Senior Managers.

- These standards are designed to maximize the benefits of a fast, accurate and well-understood CM system which will:
- Help get new products into the market faster. Reduce delivery time for customized product.
- Make happier customers because they will see the new option, change or feature they requested much quicker.
- Reduce significantly the manufacturing "bone piles" of rework and scrap material.
- Improve Bill of Material accuracy and save the corresponding material and parts costs.
- Eliminate multiple Bill of Material databases and save the costs of maintaining the databases and eliminating the risks associated with multiple databases.
- Reduce field maintenance, retrofit and repair cost.
- Reduce ERP/PLM run time. Avoid weekend computer runs that spill into Mondays.
- Make you know exactly what is non-interchangeable in each product and in every product change as required.
- Improve the understanding and communications between engineering and manufacturing and others.
- Clarify responsibilities to eliminate finger pointing.
- Save wear and tear on CM technicians, configuration managers, master schedulers, and engineers.

- Help you sort out changes that are not needed or aren't cost effective.
- Cut distribution and save paper and printing costs.
- Comply with commercial, government agency, and international standards.
- Qualify as a best-in-class producer.

These standards are especially designed to allow fast processing of releases, requests, and changes. Why is process speed so very important? These processes are "just paper/online processing," how can speed matter? Other than saying "time is money" what specifically do fast processes contribute to improved profits?

The best way to answer these questions is to ask more questions. It is a good idea to have 20-minute meetings with the people involved in the process and ask them to brainstorm why speed is important!

The questions to ask:

- How fast/slow is the current process? Perhaps 20 or 40 days just in the CM portion of the change process? (Not an unusual condition.)
- Is there more than a few hours of "hands on time" to process a change?
- How fast might this portion of the process be? Perhaps 5 days?
- What happens during the 15 to 35 unnecessary days?
- What are suppliers doing? Building items that will have to be returned, reworked or scrapped?
- What is the shop doing? Building items that will have to be reworked or scrapped?
- What is assembly and test doing? Working on items that will have to be reworked or scrapped?
- Is the line or part of the line "down"? Do we want to keep it that way for 15 to 35 extra days?
- Will the change be retrofit? Will we ship 15 to 35 more days worth of product to be retrofit in the field or factory returned?
- What if the change is a real cost reduction? Should we ship 15 to 35 days worth of product at the higher cost?
- Did the customer request the fix or feature? Should we make the customer wait 15 to 35 unnecessary days to get it?
- Is the customer site down? Would you like to be the field service person taking the heat during 15 to 35 extra days?
- What is 15 to 35 days of customer good will worth?

Yes, fast accurate and well-understood processes are key to company profitability. In one company, a five day thru-put time was achieved while the engineering and operations process times were also reduced.

A dedicated, motivated CM manager is also key to profitability and to bridging the gap between Design Engineering and Operations. For that reason, the CM manager's job description should be addressed early on in any process improvement/redesign/reengineer effort. A suggested job description is in Chapter I – Company EDC/CM Policy.

Questions and Answers

Why use the word Standards?
Because it is easier to say and because these writings include policy, procedure, forms, form instructions, flow diagrams and practices.

I'm in the process manufacturing business, not product manufacturing, will these standards help me?
Only partially and will require more editing.

How can they be "best practices" for all product manufacturers?
Each product manufacturing company will have different conditions regarding; current standards, practices, organization, attitude toward configuration management and their software applications. The standards will require editing in regards to those conditions. The standards, however, are based upon the authors work experience (5 companies), consulting experience (over 75 companies), seminar interaction (over 4,000 people from hundreds of companies), and three best-selling books in the field. That experience tells the author that the basics of CM do not change — unlike our product designs.

How can they satisfy the requirements of all commercial, military and international standards?
Allow this author to quote from one customer's letter: "_____ would like to thank you for some of the best wisdom regarding configuration management. We were recently awarded several quality awards and received our IS9001 and AS9100 certification last month. A majority of our audit revolved a great deal around design and development control. We received 100% scores in these areas resulting from techniques we used in your book. We are an old but newly managed Aerospace company serving the world. Some customers are NASA, Lockheed, Airbus, Boeing, Raytheon, and Sikorsky. All of these companies have accepted our methods and most have commented that they wish their system was as simple. We feel confident that your 'good logic' will continue to grow our company for many years to come."

Is there some redundancy in the standards?
Yes, to highlight important requirements when relative.

How are they best used?
Circumstances will vary with company size, age, extent of current documentation, company culture, management's understanding of CM and goals. A new startup product manufacturing company, with some editing, can implement much of this "system" in total. Some will want to "pick and choose" among the many standards to enhance their system. If your company has a particularly troublesome process it might take priority, but the basics should always be put in place first.

In what order should I take on the editing/implementing of the standards for my company?
Start your standards by understanding the author's "generic best of the best process" by reading Standard #47 – the Users Guide / CM Plan. You should eventually write such an "overview document / CM Plan" for your company/group. It can be given to anyone who asks; "How does the EDC/CM system work?"

If you do not have a satisfactory job description for your CM manager, edit the one found in the chapter on Company EDC / CM Policy and get management approval.

Then read and rewrite as necessary the standard on writing CM standards (#02) and then develop your acronyms standard and the definitions in Standard #48.

It is then wise to start from the "bottom up." That is, begin with basics such as the policy statement, part numbering, interchangeability, etc. (Standards 01 thru 17). Rewrite them to suit your terminology, one at a time, and obtain approval. Get some successes under your belt and then move on to harder subjectsinterchangeability).

Obtain approvals from the highest possible management level you can interest in this project.

Next, ask yourself if you are going to document current practices or attempt to improve current practices in conjunction with writing the standards. It is often best to document the current process since very often the parties involved may not even agree on what the current process is.

Flow diagram the current process and draft an improved process work flow (Release, BOM, Request or Change). Use P^3 — Principal of Planned Plagiarism — to develop your own version of that process. Write the supporting standards for that process.

How many processes are there in the "system"?
This author deems the Release, Change Request, Change as separable processes. The BOM-related standards could also be called a process but aren't easily put

into a flow diagram. The field change process is easily separated from the factory change as is done in these standards. So, the answer is: "Whatever you want them to be!" Divide and conquer when practical.

How can one show that implementation of the standards is effecting improvement?
Measure the processes as outlined in Standard ##46 and in the CM Metrics book.

What if, in some cases, I can't "sell" the author's approach?
Don't belabor any given standard if you do not get management support. Move on to the next subject. Come back to that issue later.

There are many commercial, government and international standards that one must conform to, so what effect does that have on the standards?
The principles and practices found in these standards will almost always conform to those documents. All of those documents require written standards which must be followed in practice. Some terminology editing may be necessary. Take care to read those standards yourself rather than taking an auditors word for what they say.

The CM function is sometimes found in QA or Operations, how does that affect the standards?
In one case the author found the document control function answering to the founder/CEO. The only effect on the standards is organizational editing.

How can the other folks involved better understand the guiding principles behind the standards?
Study standard #1 Policy Statement. Read the Engineering Documentation Control Handbook. It is loaded with "Rules" and "Reasons."

Why have a separate standard for requesting change than for making changes?
There are many reasons spelled out in the EDC Handbook — chief among them is that a different set of people should be involved in reviewing requests than in redesign, reviewing, approving and implementing.

What is the difference between "cost estimation" and "calculating"
The potential effective date or serial number has a significant effect on the cost. Once the effectivity date or serial is estimated, the cost can be calculated, and the potential cost will help determine whether or not the change should be made at all.

Why is the Users Guide also called a CM Plan?
Some military standards require a CM Plan. But this author thinks the better term is Users Guide.

Will my work last forever?
Expect that any standard may be revised/refined frequently and make provisions for doing that. After all, any process designed by man can be improved by man!

Should I expect the job to be easy?
Nothing in this world that is truly worthwhile is easy.

Part I

COMPANY EDC/CM POLICY

Definition

To define the company policies toward the engineering documentation control/configuration management, the CM organization and management with the rest of the company.

1

Policy Statement

Purpose

- The basic "raw materials" for any product manufacturing company are Tools (machines, molds, software, etc.), People (and the processes they devise), Money and a Product (embodied in drawings and specifications). Thus, the need to control and streamline the product documentation processes must be considered strategic planning for ________________ company.
- To assure the lowest total product life-cycle cost as well as fast and accurate well-understood Configuration Management/ Engineering Documentation Control – Product & Documentation Release, Change Request, Design Change, and Bill of Material processes. See page 5 for "Why Process Speed Is Important."
- To attain the benefits listed on page 5 of this standard.

Applicability

- Applies to the design functions and other functions served by this configuration management organization.

Policy/Practice

- To have an organized, fast, accurate and consistent process for controlling the configuration of hardware and embedded software products.
- Shall consist of planning, control, identification, traceability, and reporting.
- Shall consist of Release, Change Request, Change Control and Bill of Material processes.

Procedure

- Not applicable

Primary Responsibility

DOCUMENT CONTROL MANAGER:

- Design and documentation of the CM processes by flow diagram, form, form instruction, procedures, standards and further policy as required.
- To manage the processes and to report to senior staff as to the volume, speed, and accuracy of the processes.
- To educate and train those involved on the CM standards.

VP of ENGINEERING:

- To assure that the CM function has the necessary resources, training and authority to perform its functions.

Authorization

Authorization President/GM

Why Process Speed Is Important

- How can the process speed be important? These processes are "just paper/online processing," how can speed matter? Other than saying "time is money" what specifically in fast processes contribute to improved profits?
- The best way to answer these questions is to ask more questions? It is a good idea to have 20-minute meetings with the people involved in the process and ask them to brainstorm why speed is important!
- The questions to ask:
  - How fast/slow is the current process? Perhaps 20 or 40 days?
  - How fast might the process be? Perhaps 5 days? Is there more than a few hours of "hands on time" to process a change?
  - What happens during the 15 to 35 unnecessary days?
  - What are suppliers doing? Building items that will have to be returned, reworked or scrapped?
  - What is the shop doing? Building items that will have to be reworked or scrapped?
  - What is assembly and test doing? Working on items that will have to be reworked/retested or scrapped?
  - Is the line or part of the line "down"? Do we want to keep it that way for 15 to 35 extra days?
  - Will the change be retrofit? Will we ship 15 to 35 more days worth of product to be retrofit in the field or factory returned?
  - What if the change is a real cost reduction? Should we ship 15 to 35 days worth of product at the higher cost?
  - Did the customer request the fix or feature? Should we make the customer wait 15 to 35 unnecessary days to get it?
  - Is the site down? Would you like to be the field service person taking the heat during 15 to 35 extra days?
  - What is 15 to 35 days of customer good will worth?

Benefits of a Fast, Accurate and Well-Understood CM System

- Help get new products into the market faster. Reduce delivery time for customized product.
- Happier customers because they will see the new option, change or feature they requested much quicker.
- Reduce significantly the manufacturing "bone piles" of rework and scrap material.
- Improve Bill of Material accuracy and save the corresponding material and parts costs.

- Eliminate multiple Bill of Material databases and save the costs of maintaining the databases and eliminating the risks associated with multiple databases.
- Reduce field maintenance, retrofit and repair cost.
- Reduce ERP/PLM run time. Avoid weekend runs that spill into Mondays.
- Know exactly what is non-interchangeable in each product or every product change as required.
- Improve the understanding and communications between engineering and manufacturing and others.
- Clarify responsibilities to eliminate finger pointing.
- Save wear and tear on CM technicians, configuration managers, master schedulers, and engineers.
- Sort out changes that are not needed or aren't cost effective. Cut distribution and save paper and printing costs.
- Comply with commercial, government agency, and international standards.
- Qualify as a best-in-class producer.

EDC/CM Manager Job Description

Responsibilities:

- Owner of the configuration management processes and standards for this (company or division). These processes include the release of new items, BOM and item master files, PLM / PDM files, design change requests, design change control, change cost estimating and field change control.
- Owner of the engineering documentation control process and standards manual.
- Assure that the processes between departments are documented in form, form instruction, policy, flow diagrams and standard practices as necessary.
- Assures that these standards and CM principles are well understood by conducting the necessary training for those involved in and affected by the system.
- Control the master design documents after the appropriate point of release (pilot or production), including all digital CAD, word, PLM / PDM and hard copy master files.
- Accurate incorporation of all changes into the master documents after the changes are appropriately signed.
- Monitor all design change requests, route to the responsible engineer and/or the request review team (VPs or their directors). Route responses to the requester, resolve disputes and chair request team meetings.
- Maintain the Request and Change Action Items Lists and distribute on a same day basis.
- Control all design changes. Assure proper conformance to standards. Know the actual effectivity of every change and produce such status accounting (traceability) reports as may be required or know where to find that information. Chair change team meetings.
- Measure and report on the processes volume, thru-put time and process quality. Assure that process time and quality are continuously improving. Report to senior management at least monthly. Report to other involved managers and key people weekly.
- Assign and control all (part numbers and/or document numbers), change numbers, request numbers and document revision levels.

- Owner of the engineering data in the BOM (MRP or ERP) part master and parent component files (parts lists) and screens. Responsible for BOM accuracy with regard to design data elements.
- Know what changes are to be retrofit / affect the field and what units were changed in the field / where to find that information readily.
- Monitor all engineering design documentation for quality and adherence to standards.

Optional Responsibilities:

- Assure that an off-site emergency backup file is kept current for all design documentation.
- Control all the CAD/ERP/PDM/PLM/CM software application seats and systems. Assure maximum up time.
- Maintain an engineering library with appropriate supplier manuals, outside organization specifications, books, etc., for reference by any responsible engineer.
- Manage the component engineer(s).

Not Responsible for:

- Filing or control of manufacturing, service, quality or sales technical documentation.
- Input of non-engineering data to the item master file in MRP or ERP.
- Input to or maintenance of the approved manufacturers list.

Part II
BASIC STANDARDS

Definition

The general policies required of any product manufacturing company before implementing best-in-class processes for Release, Bill of Material, Request, Change and Field Change control.

2
Writing CM Standards

Purpose

- Briefly state the purpose of the standard. "To meet ISO 9000 requirements" is one reason for writing standards. Better reasons for writing standards are:
  1. Provide consistency in the day-to-day process operation.
  2. Furnish a basis for training people.
  3. Provide a base for improvement.
  4. Standardize the usual or normal expectations (not exceptions).
  5. Specify who may take exception to the normal expectation.

Applicability

- State the limiting aspects and the scope of the standard; example, "Engineering Change Request and closely related matters."
- Include elements of the processes which are cross-departmental. Do not include department instructions since they are/should be subject to rapid change and aren't cross-departmental business (nor are they the business of ISO certifiers).
- The standard shall not contain a separate section for cross-references to other related or associated standards. Such practice creates a "web" that is very difficult to both originate and to keep up to date. The related standards can be referred to by title in the text where they are important.

Policy/Practice

- One subject/topic should be included in one standard. Use of the word "and" in the title may indicate that separate standards should be written.
- Do not try to cover all the unique situations that have ever occurred. Leave exceptional circumstances to be handled by exception. This will normally keep the length of most standards to one to three pages. This makes them easier to read, understand and to train the affected people.

- List the statements about the subject/title that are important and are "value added" to the company. Do not try to address infrequently occurring conditions. Leave those anomalies to be handled by exceptions to the standard.
- Include elements of the processes that are critical to the interdepartmental function of those processes.
- All EDC standards will be written in the form and format herein used and described. Each shall contain the heading and subtitles: Title, Number, Date, Page of Pages, Purpose, Applicability, Policy/Practice, Procedure, Primary Responsibility, and Authorization. When not applicable, enter "NA".
- Keep sentences short. Use "bullets" as opposed to paragraph numbering. Informal polls have indicated that this method is considered friendlier than the government method of paragraph numbering (3.4.1, etc.)
- All standard numbers and form numbers shall be assigned by CM. All standard and form version dates shall be assigned by CM, unless that standard states otherwise.
- Page numbering: Normally attach referenced documents as done in the Policy Statement (01). The standard on Mark Up of Design Documents (#46) might reference a set of example marked-up docs that are displayed in CM — then those referenced documents need not be included in that standard page numbering. This would allow for those referenced documents to be separately maintained by CM.
- A new version date will be assigned when changes to a standard are made. The latest changes are/may be highlighted. All pages of that standard will be updated to the latest version date.
- Each form shall have a form instruction standard. The instruction will be "find numbered" in order to associate the instruction to the form. All forms shall have a unique date in the lower left corner — that date is the latest version of the form. An online form and form instruction (displayed when the curser is placed on a form "box") is an acceptable method.
- Online forms should be designed with the goal of keeping the printed form to a one 8 ½ × 11 page because people do make hard copies and thus we can save paper.

Procedure

- Procedure shall cover the process method required to describe interfaces between departments. Internal department practices

3

CM Requirements for Drafting Standards

Purpose

- To specify those EDC/CM requirements for design document "Drafting Standards." If the Drafting Standards already exist, they must be modified by this standard or rewritten to comply.

Applicability

- All design documents (drawings, specs, code, etc.) to be released by this CM function.

Policy/Practice

- The Company Drafting Standards shall be prepared using ANSI Y 14.x as the primary guide. They shall be written with as few requirements and pages as practical. All the following requirements will be reflected in those standards.
- They shall contain as few print formats as possible. Only A, B & C print sizes will be allowed.
- Each format will have only one place on each page for noting the latest revision level of the document.
- Multiple pages of a document will be kept at the same revision level unless Doc Control agrees to an exception to the rule.
- All formats will have a drawing revision block on the first page with the following headings:
  Rev   Date   Rev Description   ECO # Sign
- An end-product nameplate drawing will be required for each product. It will contain the following information and no other information: Serial Number, Product/Model Number, Part Number (top level), Company Logo, and Made in USA. (Serial/Date/Mod/Series Code/"Rev"/Batch # Software Release/Version Number — pick your applicable information.)
- The Serial Number (etc.) will be used for warrantee control as well as tracing non-interchangeable changes.
- The Product/Model Number will not exceed ___ characters. It will be assigned and used by Marketing for sales purposes. Since this

number is not specific as to the product's configuration, the part number will be used on the nameplate and for ordering product from the factory.

- Products with more than "x" features/options will be prepared using a Modular BOM and tabulated "drawing" per the Engineering Documentation Control Handbook.
- Sub-assemblies will not have a nameplate or be serialized without the agreement of Production Control, Service, and CM.
- Any information or data placed in the margin of a drawing shall not be under change control.
- The item description placed in the title block of the design documents will be done by the naming standard found in the Class Coding/Naming Conventions/Group Technology Standard #15. The design document will be the sole source of input of design data to the ERP or PLM.
- When the company has the capability to produce an item, its documentation will be prepared to "make." When the company does not have the capability to produce an item, its documentation will be prepared to "buy." This will normally be a Specification Control Drawing. This specification drawing will outline the critical parameters to the company applications. The Part Number will be placed in the document title block (see Part Numbers Standard).Any bar code or other labels to be left on the product at shipment must be documented by a design drawing.

Procedure

- None required.

Primary Responsibility

- The Design Drafting Department will prepare the drafting standards and maintain those in a current condition. The drafting standards must conform to this document.
- CM will keep this standard up to date.
- CM shall train all existing and new engineers and designers as to the requirements of this standard.

Authorization

Authorization Engineering Services Director

4
Technical Document Groups

Purpose

- To group technical documents and define basic control policies.

Applicability

- All technical documentation in the ____________ company/division/group supported by this Doc Control function.

Policy/Practice

- Technical documentation is divided into four categories as follows:

  - **DESIGN DOCUMENTS**

    Product Spec
    Customer Spec
    Design Spec
    Test Spec
    Part Drawing
    Silk-Screen Drawing
    PCB Artwork
    Cable / Harness Drawing
    Wiring From-To List
    Prom Truth Table
    Embedded Software Spec
    Process Spec
    Material Spec
    Spec Control Drawing
    Source Control Drawing
    Assembly Drawing & Parts List
    Logic Diagram
    Pad Master / Solder Master
    Schematics
    Software Program
    PROM / EPROM Spec

  - **MANUFACTURING PROCESS DOCUMENTS**

    Routing / Process
    Illustrations for Process
    Tool / Fixture Drawing
    Inspection Process
    CAM Programs
    Test Equip Drawing

  - **SUPPORT DOCUMENTS**

    Field Instruction/Bulletin/Kit
    Illustrated Parts Catalog
    Product Description Manual
    Parts List for Support Documents
    Spare Parts List
    Spares Kits

- **CM STANDARDS**

Forms	Flow Diagrams
Form Instruction	Procedures
Standards	Policy / Practice

- Design documents will be released (see Release chapter, filed and controlled (see Change Control Process chapter) by Engineering Doc Control/CM function.
- CM Standards will also be controlled by the CM function, not by use of the ECO Standards.
- Manufacturing Process documents will be controlled and filed in Manufacturing Engineering.
- Doc Control Standards for Manufacturing Process Documentation will be controlled by Manufacturing Engineering with the CM function agreement on the control methods.
- Support documents will be filed and controlled in Publications.
- Doc Control Standards for Support documents will be controlled by Publications and the CM function agreement with the control methods.
- The Quality Control function shall assure that the Support Docs match the product when shipped.
- Company Catalog Document will be controlled by the Sales/ Marketing function with approval of the chief engineer.
- Design changes will not be held up while affected Manufacturing or Support documents are redlined or changed. Manufacturing documentation must be redlined/revised prior to implementing the change however.
- Support documentation must be changed prior to:
  1. If shipped with the product – prior to closing the product shipping box/package.
  2. If not shipped with the product – prior to closing the design change (see Closing a Change Standard).

Procedure

- None required.

Primary Responsibility

- CM is responsible to keep this standard current.

Authorization

Authorization VP Engineering

5

Teams for All CM Processes

Purpose

To outline the company method for using cross-functional teams (sometimes called design team or concurrent engineering team) and their formation and operation. Computers, email, etc., are nice but they aren't a substitute for face-to-face discussion of new product development, product problems/suggestions or product changes. The meetings offer an opportunity to comment on documents which affect their area of responsibility (before the cognizant design engineer has put finger to keyboard or pencil to paper) and to start planning their activities. The team is not for "design by committee." The responsibility for the product design remains with the cognizant engineers.

Applicability

- All products relating to this CM Function.

Policy/Practice

- In general, encourage attendance by any function that uses design documentation.
- Attendance at the team meetings does not mean signing privileges. Each function that is affected by a design change has the right to stop a release or change through their representative that does sign (see Signatures standard).
- All functions listed on the Change Form as "affected/impacted", should send a function representative to every team meeting. Those functions are (edit to suit):
  1. UL/CAS/Other Agency Representative
  2. Component/Software/Firmware Engineer
  3. Cognizant Design Engineer/Doc Control/CM
  4. Tooling/Manufacturing Engineer/Test Engineer
  5. Production Process/ME/Production Manager
  6. Materials/Production Control/Suppliers/Purchasing
  7. Publications/Service/Quality Assurance/Marketing
- The following functions must be represented:
  1. Planner/Planner-Buyer

2. Design Cognizant Engineer for their release, request or change on the agenda
3. CM
4. ME/IE/Cost Estimator
5. Service

- Representatives may be management or a key person. A backup should be designated for each representative. The chair should maintain (preferably online) a list of functions, representatives and alternates. Any management person is welcome to attend any meeting.
- The cognizant design manager (or CM manager) will chair the new product release meetings:
- The meetings will be for one hour at least biweekly for new designs and major redesigns. They will be held in the engineering conference room at 11:00 am (to ASSURE ONE HOUR LIMIT).
- "Minutes" will not be kept but an action items list will be kept by CM – showing action items, description of the action needed, who it is assigned to, their commitment date for resolution.
- Action items list will be published by the chair by 4:00 the same day of the meeting via email.
- The CM manager will normally chair the **Change Request Meetings** of high level engineering, operations and service VPs or directors:

    1. Change requests (see Change Request Policy Standard) will be reviewed with a critical eye in biweekly short meetings.
    2. "Minutes" will not be kept but an action items list will be kept by CM showing action items, description of the action needed, who assigned to, their commitment date for resolution.
    3. Action items will be published by CM the same day.

- The CM manager will normally chair the **Change Control Meetings**:

    1. The meetings will be for one half hour each and every workday in a conference room without chairs.
    2. Organization by design group or product lines or some combination may be best. The important thing is to keep the meetings short and hold them more frequently.
    3. "Minutes" will not be kept but an action items list will be kept showing action item, description of the action needed, who assigned to, their commitment date for resolution.

4. The action items list will be published by the CM manager the same day via email.

- The chair will first call for discussion of the action items. Then ask for cognizant engineer(s) to briefly present new design plans, change requests (optionally done by the CM manager) and change orders. See applicable standards for other uses of the team in release, request and change processes.
- All test results will be brought before the team as quickly as reports are available.
- When lengthy discussions seem to be developing, the chair must assign action as appropriate, including "outside" discussion/meeting if necessary to resolve an issue. Items not due per the action dates will normally not be discussed.
- The cognizant engineer shall obtain signatures from other technical people as required by standards. A computer will be available at the team meeting for online signatures.

Procedure

- See flow diagrams.

Primary Responsibility

- The engineering services director is responsible for keeping this standard current.

Authorization

Authorization VP Engineering

6

Cognizant Engineers

Purpose

- To delineate, chart, and define the responsibilities of design engineers and others regarding product design and its documentation.

Applicability

- Applies to all engineering documentation control processes controlled by this CM function.

Policy/Practice

- Cognizance is responsibility and signature authority if affected.
- This authority is to be delegated to the lowest practical level. Managers or project leaders often are bottlenecks. Managers can and should interject themselves into the process without signature authority as required — depending upon the skill of the cognizant engineer. Empower the key professionals. How does the engineering manager or project manager decide who to give the request/problem/suggestion to? That is the cognizant engineer.
- Responsibilities are defined as follows:
  1. Cognizant Design Engineer – The person who requests are given to and changes accepted from. Responsible for the design of the part, assembly, product, or system assigned. Consults with other design engineers as necessary – UL, Component, System, Project, etc. Assures that all other cognizant engineers who use an affected item agree with the change to the item.
  2. Configuration Management Technician – Responsible for the operation of the CM processes and systems, BOM accuracy, etc.
  3. Manufacturing Engineer – Responsible for the design of the manufacturing process. Consults Test Engr, Industrial Engr, Receiving Inspection, Suppliers, etc., as necessary.
  4. Planner/Planner-Buyer – Responsible for planning, tracking, implementing and capturing the actual effectivity of the change.
  5. Field Support Representative – Responsible for the repair and maintenance processes. Responsible for field implementation of

specified non-interchangeable changes. Responsible for implementation of certain changes in returned goods.

6. Quality Assurance Engineer – Responsible for audit of the above processes and reporting to the management as to their effectiveness.

- CM shall maintain a chart of cognizant engineers (and each alternate) by product, assembly, part, customer, etc. (pick one). They will update the list at least once a month by asking the appropriate management people to mark up the list.
- Cognizant Engineer List – example:

Item	Design Engr	ME	PC	CM	QE
Frame	C Brown	J Eads	F Man	E Swift	J Braun
Motor	J Walker	J Eads	C Swift	M Wait	J Braun
Component	A Bunker	E Carn	C Swift	E Swift	C Katt

Procedure

- None required.

Primary Responsibility

- The CM manager will maintain the Cognizant Engineer List and this standard. It will be reviewed/updated by the applicable management every month.

Authorization

Authorization — Engineering Services Director

7
Part Numbers

Purpose

- To define the attributes of the company part number.

Applicability

- All items and their documents to be released by this CM function.

Policy/Practice

- The part number will have the following characteristics:
  1. Document number embedded in the part number (avoids cross-references).
  2. Tabulation ability (see below).
  3. Number of digits in length expected to last 50 to 100 years.
  4. Minimum significance (see the article attached).
  5. Compliments a predetermined naming convention (title and description) or class code in order to avoid reinventing the wheel, to find substitutes, etc. (not part of the part number).
  6. On non-interchangeable (part number) changes, only the tab number will change.
- All part numbers will be eight digits numeric according to the following convention:

N N N N N N - N N (N = Numeric)

DOCUMENT #

TAB #

PART NUMBER

- The dash is shown only for example, not to be used on the document or in the database (avoids unnecessary keystrokes).
- The tab number is a convention used in the drawing title block and in the database as follows:

1. If the document does not represent a physical item (document only) then use . . . "00"
2. If the document is tabulated (variations of nearly identical items) use . . . . "XX"
3. The first part delineated on the document will always (tabulated or not) be . . . . . "01"

- A crude but sometimes adequate form of class coding may be used by a couple of alpha/numeric prefix characters — if necessary to "sell" the above "ideal" PN. A separate class code is preferred.

Procedure

- CM assigns all part numbers.

Primary Responsibility

- CM will keep this standard up to date.

Authorization

Authorization	VP Engineering

Significance in Part Numbers

There are zealots on both sides of the issue concerning significance in part numbers. Most product manufacturing enterprises, at one time or another, have this controversy. Every new start up has this issue. The issue permeates the operation. It crosses all functional lines. Some say that significance (smart number) is the only way to go. Others say that non-significance (dumb number) is the only way to go. In this writer's opinion, either extreme is generally wrong. A generalization for items from the top level to the bottom level of your structure cannot be made to stand the test of logic. Also, any discussion of part number design without inclusion of interchangeability consideration is ludicrous.

While believing in minimum significance, this writer would submit that at the top level — your product level — the smart number might make very good sense for products with few features and options. Consider the company wherein the salesman structures a smart number from a catalog when determining what features and options the customer wishes (catalog configured). Should we take the order by catalog number and have Order Entry convert it into a factory part number? Won't the conversion inject a possibility of error into the process? When the conversion error occurs (and it will), the customer will receive something different than they wanted — an intolerable situation. In this circumstance why not use the configured catalog number as the top level/end item number in the system. If maintaining a catalog works for you, look seriously at using that smart catalog number as your end item number. The same logic might apply to other "configurable assemblies" within the product that are separate line items on your sales order.

If your catalog is too difficult to maintain because of numerous/added/changing features, then a minimum significant number for the top level (and configurable assemblies) would be in order. A modular bill of material or configurator module at the top level could be the answer.

Before exploring lower level items let's ask ourselves, "Why significance at all?" The temptation to use a significant part number is high. The significant part number helps us to find similar parts. If we don't have significance in the part number, how do we search to find similar parts? How do we avoid reinventing the wheel? How do we find a temporary substitute for a part shortage? How do we know which parts might be manufactured in a single cell? Or perhaps your history has produced several part numbers for interchangeable items (17 at one company) and you want to standardize.

A group technology or class code system (smart numbers) outside the part number is sometimes the answer. With the power of computer word searches, an intelligent description field or "naming convention" (another form of class coding) might be the answer. With the advent of low-cost computing, it is probably better to set up a database with a separate field for each characteristic that might have otherwise been put into a significant part number. Then searches can easily be done on those fields. If you are a new company, or if your smart number has broken down, or you are merging several operations into one, a good rule to follow would be:

Rule: Put very little significance into the part number with the possible exception of end items with few features and options.

Reason: Because significant numbering systems tend to breakdown. No matter how good you are at anticipating the number of digits you will set aside for a given characteristic, at some point it won't be enough. There are many pros and cons to each practice, but the potential breakdown of the significance is the final determiner.

The next question that should be addressed is whether or not a separate drawing or document number is required. Many companies have established (legacy) systems with separate numbers on a part and its document. This requires cross-references to manage to navigate from one to the other. If you have this system and it works for you, don't change it. Again, however, if you are a new company, your smart number has broken down, or you are merging several operations into one and seeking a single part numbering system, a good rule to follow would be:

Rule: Embed the document number in the part number.

Reason: Cross-referencing takes an instant at best. Multiply that instant times all the people that will have to cross-reference (both ways) for the rest of their lives and their replacements' lives.

This also tends to tear down that wall that often develops between engineering and manufacturing. They both would use the same number — what a simple but powerful communication bridge! Face it, embedding the document number in the part number is a form of significance even when it is a serial/dumb number.

Last but not least, the question must be asked: "What if we change an item non-interchangeably after release?" The part number should change. Ever since Eli Whitney, non-interchangeable changes meant a different part number (in one or more characters). Some operations say, "We don't change part numbers — only document revisions!" This is a very expensive policy. It works fine when the product is young and most changes are required to meet specifications. But what about the mature product wherein most changes are made to reduce cost? You certainly don't want to retrofit cost reductions, so we better find the best way to accomplish part number changes.

Think of all those folks that memorize part numbers — and there are lots of them. Are we going to ask them to rip the old number out of their memory and insert a completely new one? The answer is to attach a "tab" or "dash" number on the document number. One or two digits should do most operations quite nicely. These digits can also serve to document similar items on the same document — a basic form of class coding. Thus, on a non-interchangeable change the number would change from "01" to "02" and a new document is not needed. It also allows the document revision level to be set aside for document only and interchangeable changes.

This writer would submit that a part number without a tab/dash is like a ship without a rudder! Thus, the "ideal part number" would be a format as follows: YYYYYZZ, where "Y" is a sequential (dumb) document number and where "Z" is a dash/tab number ("smart") — all numeric. Together they make up a part number. Thus, another rule:

Rule: Always tabulate the part number. If you have an existing system that does not include this feature, add it as soon as practical.

Reason: Saves some labor to prepare documents and to revise them. Allows changing the tab upon non-interchangeable change and thus is friendly to those many people who memorize part numbers. Takes away an excuse for not changing part numbers on non-interchangeable changes and serves as a "version number" for repair and refurbishment.

If a company has a couple of digits of class code in their current part number it might be serving them quite well and thus shouldn't be tampered with. If you have a class code prefix of a couple of digits of alpha or numeric it would be acceptable to keep it, especially if that allows you to "sell" the new number format. Thus, the number would be: XXYYYYYZZ, where "X" is the legacy class code. It would also allow the base number to stay the same for the board, artwork, and assembly — a good aid for the electrical types.

Thus, the ideal part number is significant by virtue of embedding the document number and having a tab suffix. As stated before, the significant part number may have proper use at the top level in some companies. It is therefore this writer's opinion that a wise part numbering system may not be either totally significant or totally non-significant, but have "minimum significance."

8

Quantity and Units of Measure

Purpose

- To define the acceptable units of measure, avoid separate "engineering" units and "purchasing" units, and define proper units of measure usage.

Applicability

- All part information and parts list files released through this CM function.

Policy/Practice

- The quantity shown on the parts list will be the quantity per that specific assembly.
- All pilot and production released assemblies must have a quantity specified on the parts list even if the item is designated as a "floor stock" item. These items may have an entry in the quantity field of "AR" (As Required) during pilot production but must be exactly quantified before production release. To obtain accurate product cost (see next bullet).
- If the item is included in the product cost as part of the material burden/overhead it may be "quantity AR" and quantified in the assembly process instruction.
- The parts list will specify an acceptable unit of measure per the following list:
  1. **EA** (each) – All items that are normally purchased on a piece basis.
  2. **FT** (feet) – All items that are normally purchased by the foot on rolls, spools or reels such as wire, shrink tubing, hose, etc.
  3. **PR** (pair) – All items which are purchased and issued in sets of two. Such items are normally matched or complimentary in nature – latches, hinges, etc.
  4. **Ref** (reference) – Indicates that the part number is associated only with a document that does not depict a physical item.

5. Etc. (add if needed).

- Only design reference documents will be included in the Item Master or the Parts List (exclude ME and QE referenced docs).
- Items used in varying quantities can be assigned an "actual" quantity by determining an average quantity used. Example: Shims are required to properly adjust the skew on the capstan motor. Figure the average requirements and round up to the nearest whole.
- Items requiring a portion of a foot/pound/etc. may be expressed in a decimal as small as .001 (check ERP system).
- Tool life may also be expressed in decimals to track the tool's end of life.
- ERP systems may not offer a "where used" ability if the quantity is "Ref" (check your ERP).

Procedure

- None required.

Primary Responsibility

- Production Control Department will be responsible for keeping this standard current.

Authorization

Authorization VP Engineering VP Operations

9

Bills of Material

Purpose

- To define BOM content and structure.

Applicability

- All parts lists released by this CM function.

Policy/Practice

- The parts list shall contain all product design related parts, assemblies, design drawings, specs, code, packaging/boxing material, labels and publications and literature shipped with the product.
- The BOM Item Master data elements which are Engineering's responsibility are: Find Number (tie to the pictorial), Part Number, Item Description, Weight (optional), latest drawing Revision Level, Unit of Measure, Quantity per assembly and the ECO # releasing or changing. (Include cognizant engineer if assigned by part.)
- Engineering data elements will be maintained at 100% accuracy by CM.
- All other data elements will be input by the responsible function – Labor Rates and Overhead Rates by accounting, Labor Hours by IE, Lead Time by planner, etc.
- **One structure/one file/one database will be our goal.** Normally the parts list (and changes theirto) will not be input to the CAD, not placed on assembly drawings, but are input once to the ERP file.
- The ERP shall be programmed to produce a parts list containing only those elements that Engineering wishes (see EDC Handbook). This parts list will be titled "Engineering Controlled Parts List" and double spaced. (Used as a "controlled document" to mark up for change.)
- The BOM structure will be jointly developed by Engineering and Manufacturing. The following guidelines will be used:
  1. Fewest possible assembly levels.

2. Modular structure if features and options (see EDC Handbook).
3. Standard/Common Assemblies (used in multiple products).
4. Structured to match the manufacturing processing plan but no assembly shall be added solely to allow "stocking" of an assembly.

Procedure

- None required.

Primary Responsibility

- CM will maintain this standard up to date.

Authorization

____

Authorization VP Engineering

10

Approved Manufacturers List (AML)

Purpose

- To specify the method of controlling the acceptable manufacturer(s) for each purchased item (often called AVL – Approved Vendor List; or QVL – Qualified Vendor List).

Applicability

- All new purchased items released by this CM function.

Policy/Practice

- The control of the source for purchased items shall be done solely by an Approved Manufacturers List (AML). This list will be maintained for all purchased items in the ERP Item Master file.
- Engineering shall furnish the first acceptable manufacturer for each new design item to be purchased. The buyers must be consulted in this effort. The first manufacturer will be named on the form releasing that item, not on the item's document.
- It is desirable for Purchasing to add other manufacturers as appropriate. At least one additional manufacturer will be added to the AML in the first year after production release.
- Engineering, Purchasing and Quality Assurance must agree on those manufacturers as acceptable sources.
- It will contain the manufacturer name, not wholesalers, retailers or other sources.
- The supply chain may not purchase parts made from manufacturers who are not on the list for that item. They may purchase the item from any supplier (Radio Shack, for example) as long as the item was made by the approved manufacturer(s).
- The AML input and maintenance will be the responsibility of Quality Assurance. They will not use the design change form but rather design a separate AML change form/formatted email for control. That AML form will not be used for design/document changes. Forms will be filed by QA.

- Quality Assurance will obtain Design Engineering and Purchasing Agreement with all changes to the AML. Option: Manufacturing/ Industrial Engineering may represent Purchasing.

Procedure

- Procedure according to the AML change form as specified by the quality assurance function.

Primary Responsibility

- Quality engineering manager will maintain this standard in a current condition.
- Quality assurance director may take exception to this standard by outlining the reason in the AML change form.

Authorization

Authorization VP Quality Assurance

11

Deviations

Purpose

- To define the circumstances wherein a deviation from existing design criteria/documentation is acceptable.

Alternate Method

- Do not allow deviations at all. If the engineer won't write an ECO to the item design then rework to print, return to supplier or scrap the item (author preferred method).

Applicability

- All deviations to design documentation as released or revised by this CM function.

Policy/Practice

- Use only for a temporary departure from the design drawings and specifications — thus there is no need to change the design documents. If the design needs to change use the Change Form and procedure wherein a deviation is not allowed or required.
- The deviation shall not be used as another way to make design changes.
- Repeated deviations for the same problem usually means:
  1. The person designated to be responsible for "clearing" the earlier deviation did not/could not accomplish the task of fixing the root cause problem. That person should be consulted prior to writing another deviation for the same problem.
  2. The producer of the items involved does not understand the design criteria or does not have the latest criteria.
  3. A design change may be required. If so, go directly to the Change Form, do not write a deviation.
- All deviations will be documented on the form shown on page 5 and signed by QA, ME and the cognizant design engineer.

- Quality Assurance shall control the writing and approval of all deviation forms.
- The deviation number of all deviations shall be posted in the revision block of the applicable drawing by CM (without writing an ECO) and a copy filed in CM. (Traceability – allows engineers in future troubleshooting to readily see the deviations.)
- QA will place a form instruction online and/or on the back of the form (see the EDC Handbook).

Procedure

- To be specified by QA.

Alternate Method

Do not allow deviations at all. If the engineer won't write an ECO to the item design then rework to print, return to supplier or scrap the item (author preferred method).

Primary Responsibility

- Quality Assurance will keep this standard current.
- Quality assurance director may take exception to this standard by noting in the deviation what was done and why.

Authorization

Authorization		CM Manager

Deviation Form

Deviation Authorization	**(company name)**	DA#
Originator Name	Dept	Date

Reason for Deviation:

Alternate Method

Do not allow deviations at all. If the engineer won't write an ECO to the item design then rework to print, return to supplier or scrap the item.

Specific Time Frame or Serials to be Deviated

Product(s) Affected

P N Affected   Documented Condition   Deviated Condition

Name and Organization Responsible for Clearing This Deviation

---

Has This Condition Been Deviated Before?   Yes ☐   No ☐

If So: DA #   Date   Time Frame or Units

QA Name   Sign   Date

Mfg Engr Name   Sign   Date

Cog Engr Name   Sign   Date

Doc Control Name   Date Filed

Form # _____   If a design change is required, go directly to the ECO, do not write a deviation.

12
Spares List

Purpose

- To define spare parts and spare assemblies and to specify what meaning that has for the company.

Applicability

- Applies to all products released by this CM function.

Policy/Practice

- Spare parts shall be defined as those parts, assemblies, and special assemblies which are subject to damage, ware or failure and will be packaged/stocked by Field Service for quick replacement.
- Parts of an inseparable item shall not be spared. Examples: Potted assembly, weldment, etc.
- To offer an item as a spare must be a conscious and careful decision agreed upon by Engineering and Field Service. Without this decision, it is implied to customers that all parts of the product are available for quick replacement — not wise or desirable.
- The company will furnish any item of any product but, if the item is not spared, the cost and delivery must be negotiated.
- "Special Spare Assemblies" are defined as assemblies that are not used in the normal production process. Example: Any item, with package and an instruction for installation/test.
- A separate "Spares BOM" will be prepared for each product and titled "Spares BOM for Product XXXX." All spare parts, assemblies and special assemblies agreed to be spared will be shown on that BOM. All spared items will thus appear in the ERP "where used" as being a spared/service item.
- Only those items in the Spares BOM will be listed/identified in the published Spares List/Manuals.
- Engineering changes during the pilot and production phase must also change the Spares List(s) if affected.

- Publications must reflect the changes in the published Spares List, including the serial number(s) at which the changes were effective.
- The published Spares List must be ready to ship with the first production unit. Subsequent changes that affect the spare parts lists must be reflected in the published Spares List in time to ship with the changed product.

Procedure

- Items to be spared must be initially identified during the pilot phase by Design Engineering. Engineering and Field Service must agree on the spares BOM before production release of the product.

Primary Responsibility

- The CM organization will be responsible for keeping this standard up to date.

Authorization

Authorization	Engineering Services Director & Field Service Director

13

Prints/Point of Use/Paperless

Purpose

- To define company policy regarding distribution and control of design documents.

Applicability

- All releases and changes to be processed by this CM function.

Policy/Practice

- Limit the distribution of documents to the "action documents" – release forms and change forms. These forms show the items released/changed and thus allow knowledge of drawing and spec availability.
- To distribute action documents online to all who need to know.
- Not to distribute released/revised copies of drawings, specifications, parts lists, etc., except on a demand basis. Those who review the action documents can see their needs and pull the required copies when and as needed (Pull not Push).
- To encourage online viewing (via CAD, PDM, ERP, etc.) as opposed to hard copy distribution, everyone in the organization has a responsibility to view rather than print whenever practical.
- Encourage use of online systems by not making any other list(s) of released or changed drawings, specifications or parts list.
- To assure prints and revisions to prints are at the points of use by establishment of a technical support function for each point of use. The technical support function will be on distribution for the action documents. Those functions and the area they support are:

1. Purchasing	Suppliers
2. Test Engineering	Test Floor
3. ME/IE	Assembly Line & Fabrication Shop
4. Quality Assurance	Receiving Inspection & Reliability Testing

- CM has the responsibility to offer fast "while you wait" print service.
- Each technical support function has the responsibility to review the releases and changes, understand the effective dates, obtain prints as necessary, place those prints at the point of use on the correct date, remove the old or down level prints and to destroy them. Ditto with online systems.

Procedure

- As needed by each Tech Support function.

Primary Responsibility

- CM has the responsibility to keep this standard up to date.

Authorization

____

Authorization President or GM or Chief Engr

14
Signatures

Purpose

- To specify which functions should review and/or sign a release, request or change, to define their responsibilities, and how functions that don't sign can stop the process with cause.

Applicability

- All releases, requests and changes processed by this Doc Control function.

Policy/Practice

- Engineering, Operations, and Service shall each have one signature. The signers named by the VP of those functions may vary by product, part of the product, etc. Service needs only sign a change if the change is to be retrofit.
- All functions on the Impacts/Affects list shall be represented by their organization representative. Examples: UL/CSA Engineer is represented by Engineering. Quality is represented by Manufacturing, Publications is represented by the Service Representative, etc.
- Any function which may be impacted by any given release or change will be on the "Impacts/Affects" list in the change form. They will therefore be aware of the change.
- Each function on the impacts list has the responsibility to review the release or change in the 24 hours allowed and to correct impacted "yes" or "no" checks on the change form if needed.
- Each function actually affected has the right to tell their representative (alternately the CM manager) not to sign the mark-ups or the form by writing specific objections. That representative must resolve the issue or not sign. Frivolous issues as judged by the CM manager shall be ignored.
- Only the cognizant design engineer and CM (or CM only) need sign "Document Only" changes.

- The signature/functions signing changes need not be the same as the signers of the original release of the documents involved. (Without this statement, ISO 9XXX will expect them to be the same people.)
- Technical functions will sign the drawings/mark-ups. Administrative functions will sign the change form. Service will sign the change form, if retrofit is specified, thus agreeing with the retrofit plan. If the "From-To" is on the change form instead of attaching mark-ups, technical functions will sign the change form.

Procedure

- CM shall keep a chart of signers and alternates (see Standard 06).
- They will present that list to the appropriate management at least once a month for editing.

Responsibility

- All applicable management has the responsibility for not allowing the signer and alternate to be absent at the same time. If both are absent, the manager must fulfill their responsibilities.
- CM manager may take exception to this standard by noting on the form why that was done.
- CM is responsible for keeping this standard up to date.

Authorization

Authorization: VP Engineering
VP Operations
VP Service

15

Class Coding/Naming Conventions/ Group Technology

Purpose

- To outline the choices available for creating an aid for engineers and configuration management to:
  1. Avoid reinventing the wheel.
  2. Find parts which are identical or may be identical but are part numbered differently (duplicate part numbers for the same part).
  3. Find part substitutes or parts that can be modified to substitute for a part that is back-ordered/short to production requirements.
  4. Find similar items that can possibly be produced in the same manufacturing "cell."
- Those wishing to find out more information about the "Group Technology" term should start by obtaining and reading an article of that title by Frederick B. Ingram, Published in the Fourth Quarter, 1982, Production and Inventory Management Journal, a publication of the APICS organization. This article was the 1981 International Student Award Program winner in the undergraduate division of APICS Educational and Research Foundation and is still significant.

Applicability

- This method (pick one) can be applied to: 1. All new item designs only, 2. All new and revised designs, 3. All active (used on items with orders in the past x months) items, etc. This decision needs to be carefully considered based upon the company history and kind of business.

Policy/Practice

- Alternatives: Pick one depending upon your needs and history. Seriously consider using the naming convention alone — without other methods.

- Significant part number. Assign meaning to all or nearly all of the digits in the part number for all items. This method has some useful application for some company's end items. Some products lend themselves to a catalog wherein the customer determines what features and options they desire and in so doing develop a significant part number or catalog number for ordering. Beyond these few useful applications, the significant part number has a common flaw that should deter its use. The assignment of significant digits tends to breakdown in time. Example: A couple of digits may be set aside for a particular characteristic and, over time, more than 99 variations are developed (see Procedure A).
- A Class Code prefix. Some companies have applied a leading two or three digits to their part number which identify items into predetermined categories. This is a crude but effective method for this need (see Procedure B).
- A separate field in the database is sometimes used in the CAD/PDM/PLM or ERP system for a Class Code as described in Procedure B. Assignment of a significant number in a separate field in a database but not part of the part number. The assignment of the number is based on criteria in a "code book." All digits are significant. A single company function needs to be the keeper of the code book.
- Some software programs are available to aid in this area. CMS (Component Supplier Management) packages are an alternative that should be explored, especially for larger companies. Web search "CMS."
- Last but not least is the use of naming conventions. This is a solution that is increasing in popularity since the ability of data processing systems has improved their "word search" ability. This alternative can easily be used alone or in conjunction with the other methods available. A specific set of rules and examples are required for best results. This is the alternative that young/startup companies should consider (see Procedure C).

Procedure

- Procedure A – Significant Part Number Digits (least desirable – see Part Numbers Standard):

  Example: AY074286C3234 Where:

  AY = Two alpha digits for the item "type."

  07 = Two numeric digits for the color of the item.

42 = Two numeric digits for the intake size.
86 = Two numeric digits for the exhaust size.
C = One alpha digit for the method of control.
3234 = Four digit numeric sequential number.

- Procedure B – Class Code alphabetical listing
  01 Adapter 03 Adhesive 05 Arm 07 Assembly (not PCB)
  09 Batteries 11 Bushing 13 Cable (connector two ends)
  15 Cable (harness - no connectors) 17 Cable (other) 19 Capacitors
  21 Castings, raw 23 Castings, processed 25 Chassis

- Procedure C – Naming Conventions – Examples (use this list when the rules are in question)
  <u>Noun Name, Modifiers, Value</u>
  Resistor,Carbon,20ohm,2%
  Strut,Wing,Left,Aluminum
  Washer,Fiber,Shoulder
  Chassis,Weldment,SS
  Wire,Insul,18gage,yellow

- Naming Conventions – Rules – Examples (write your own rules)
  1. No period after abbreviations – a standard for acceptable abbreviations is required.
  2. No space after punctuation.
  3. Most significant noun name first, then most significant modifier, etc.
  4. Items always purchased that are assemblies should not include the word "assembly" – Example: Motor, AC, Head Actuator.
  5. Made items (or sometimes made and sometimes purchased) that are assemblies should have the word "assembly" included and following the most significant noun name – Example: Head Assy (not Assy Head).
  6. For PCB's unassembled: PCB,Fab,Title,Ref Symbol
  7. For PCB's assembled: PCB,Assy,Title,Ref Symbol
  8. Do not list the used-on product as other products may use the same item — now or later.
  9. The drawing page one title block shall contain the naming convention exactly as required in the database. CM may correct titles to conform to the standard at release of the item or upon change. Subsequent pages of the drawing need only show the noun name. Data entry shall use page one of the drawing as the only source for the item's naming convention.

10. Check your ERP and CAD/PDM/PLM for digit limitations and existence of a "Description 2" field. It might be used for an expanded description but shouldn't be used for the naming convention.
11. Values may be required for some items – add the method for values of various types of parts.
12. Tabulated drawings should not have the tabulated characteristics in the naming convention. The engineers search leading to a tabulated document is the desired endpoint.
13. Supplier names should not be used in the naming convention.
14. Modify these or add other rules for your kind of business.

Primary Responsibility

- It is the responsibility of the configuration management manager to keep this standard up to date.

Authorization

Authorization	Chief Engineer	Date

16

Tabulated Documents

Purpose

- To indicate the proper method of tabulating similar item, part or component documents.
- To indicate the proper method of changing tabulated documents revision level and part numbers.

Applicability

- All design documents processed by this CM function.

Policy/Practice

- Similar items shall be tabulated (also see the Part Numbers Standard).
- Use of the tabulation document will be limited to items having one or two variable characteristics. More than two variable characteristics require the approval of the CM manager. This is necessary because the drawing or spec with three or more variable characteristics is very hard to understand.
- An "XX" will be placed after the document number in the title block of the drawing/spec. This will serve as a flag for the user of the document to know that the document is tabulated; Example: 123456XX.
- The "00" tab shall be reserved for the document itself. The first item documented on such a drawing/spec shall always be 01, the next 02, etc. Tab 99 is reserved for CM use only.
- A chart (tabulation) will normally be placed in the upper left corner of the document – showing the variable characteristics of the item with the tab number for each.

  Example:

Tab	Length "L"	Width "D"
01	2.5 inches	1.0 inches
02	3.0 inches	1.0 inches
03	3.0 inches	1.5 inches
04	4.0 inches	2.0 inches

- Fixed relationships between the tab number and the variable characteristics (such as 01 = 1" width, 02 = 2" width, etc.) shall only be done with the written approval of the CM manager.
- In the unlikely event that a non-interchangeable change is made to the item tabulated, and if a new part number is required (see the Interchangeability Standard), a new series of tabs must be added to the chart – In the above example: 05, 06, 07, 08.
- When non-interchangeable changes are made per the above example the tabs 01 thru 04 "where used" must be examined for proper use/designation under the company's "Inactive/Obsolete" standard.
- Any change made to the document after release must be done by ECO and result in an increase in the document revision level.
- The latest document revision level will be indicated in the ERP Item master file on the 00 tab only (Documents have Revs, parts don't). If the ERP requires entry of a revision level for each part, all other tabs shall be entered as rev level 99 (Drawings have Revs, parts don't).
- Purchasing and Production Control/Planning must be brought into a discussion on the use of this convention, especially if the MRP/ERP system produces Purchase Orders or Mfg Shop Orders.

Procedure

- None required.

Primary Responsibility

- The CM manager will be responsible for maintaining this standard in a current status.

Authorization

Authorization Engineering Services Director

17

Nameplate and Serial Number

Purpose

1. To require end products to contain a nameplate.
2. To require certain major assemblies to contain a nameplate.
3. To designate the data which will appear on the nameplate(s).
4. To require preparation of a nameplate drawing(s) meeting these requirements.

Applicability

- All products controlled by or to be controlled by this CM function.

Policy/Practice

- Each end product shall be identified with a nameplate placed in a prominent position on the product.
- All printing on the nameplate must be in __________ font.
- Design Engineering shall prepare a drawing indicating the placement of the required data on the nameplate, size of the font and the nameplate material.
- Design Engineering shall also include placement of the nameplate on the product in the appropriate assembly drawing or specification.
- The data which must appear on the nameplate/drawing is as follows:
  1. The company logo and address/phone number – consult with Management and Marketing.
  2. Model Name/Number as assigned by Marketing/Sales (typically non-specific as to configuration of the product).
  3. Part Number of the top level in the product structure (product configuration specific but not affected by design changes).
  4. A serial number for tracing and tracking non-interchangeable changes. Configuration Management shall control the assignment of SNs. They shall begin with 001 for pilot units and begin with 100 for production units. If the same product is to be built

in more than one plant a prefix alpha character will be assigned by CM to designate the building plant. CM may give Production Control a block of SNs for their assignment of a unique number to each product.

5. The assignment of individual SNs will be done before the product is tested so that test data may be traced to SN.
6. Production Control shall be responsible for assuring that the ship date of each SN is available to all who need to know, including Engineering, Field Service, etc.
7. The SN ship date shall be the date used for warrantee purposes.

- Non-interchangeable changes will be traced to the serial number(s) containing the change. Example: SNs 1070, 1072, 1073, 1076, 1078 and higher. Thus, the company will know exactly which units contain the change and which ones do not.
- Instead of tracing changes to serial numbers, see Actual Effectivity Tracking Standard for other methods of tracking changes and for other related info.
- Nameplate drawings shall be approved by the cognizant engineer plus ____________ (Sales, Manufacturing and Service Representative — take your pick of one — or more).

Procedure

- None required.

Primary Responsibility

- The CM manager will be responsible for maintaining this standard in a current status.

Authorization

____

Authorization General Manager

Part III

PRODUCT AND DOCUMENT RELEASE PROCESS

Definition:

Document the CM processes for Product Development's handoff of new design/documents to Operations for Pilot Production and then to Production.

18

Release Policy

Purpose

- To specify the policy and practices to be followed during the process of releasing parts, assemblies, products and their documents from Design Engineering to Operations.

Applicability

- All release actions to be performed by this CM function.

Policy/Practice

- The Change Form shall be used to release new items as well as to make design changes.
- New parts required for a design change shall be released as part of that change order.
- A physical item is released when its document is released.
- All new designs and "spin off" designs shall go through the normal company phases as spelled out in the Phase Release Standard. Example: If there are three phases – Design/Develop, Pilot and Production; then there are two releases, one to "release to pilot" and another to "release to production," for each new part, assembly, design document and product.
- Part releases need not and should not be "queued up" for release as an assembly or product.
- All parts must be released to the subject phase before the assembly can be released for that phase. Examples: All the parts/documents of an assembly must be pilot released before the assembly can be pilot released. All the parts/documents of an assembly must be released for production prior to the assembly being production released.
- Parts shall be released in lead time to produce. The item's "used on" assembly need not be released. The structure need not be agreed upon. Long lead parts can be structured to the top-level product until the assembly structure is agreed upon.

- Parts already released for an existing product need not be re-released for a new product using those parts.
- The ERP must be coded item by item with a "status" which indicates what phase the part is acceptable (valid) for. Example: Status code E = Engineering Design and Development Phase; S = Pilot/Sample Phase; P = Production Phase (see Phase Release Standard).
- When viewing the design documents the phase that the drawing/item is acceptable (valid for) is indicated by (also see Phase Release Standard):

  Rev blank (date controlled) = Engineering Design and Development

  Rev numeric = Pilot Phase

  Rev alpha = Production Phase
- By taking the above two steps, the Release Phase that the item is valid for can be seen either by viewing the online BOM system or the document.
- All rules for release apply as shown in the Phase Release Standard.
- All critical components must be qualified (tested) before production release.
- Production release of a product can only be affected after satisfying all requirements of the Pilot Phase and all design documents must be released for production.
- When applicable, all products must be approved and listed by any/all certifying agencies prior to production release.
- When pilot units are shipped to a customer — even to a "friendly beta site" — they shall be accompanied by a Deviation explaining all the criteria not met for production units.
- The Change Form authorizing a release will be distributed online to any who need to know.
- Hard copy of the documents being released will not be distributed to anyone. The Impacted/Affected functions will use the release change form to determine their need for hard copies and obtain them when and if needed (pull not push).
- Any person who believes that a particular release is urgent may personally expedite/"hand-carry" the form through the process. CM will not hand-carry.
- Problems encountered during the release process should be brought to the attention of the CM technician for resolution.

Procedure

- None required (see the Change Form for release form instructions).

Primary Responsibility

- The engineering services director shall be responsible for keeping this standard up to date.

Authorization

Authorization	VP Engineering	VP Operations

19

Team in Release

Purpose

- To outline the function of the Release Team in the release of parts, assemblies, products and their documentation. Identifying and resolving design or document issues. Tracking the progress of the development of the product, its parts and documentation through the normal company phases.

Applicability

- All items to be released by this CM function.

Policy/Practice

- Assure that the Release Policy Standard is followed and review all new product designs by review of plans, layouts, 3-Ds, drawings, etc.
- Continually update and track the expected product cost.
- Facilitate the release of new items in as short a time frame as practical.
- Determine the number of pilot units to be built.
- Release all items in lead time to manufacture.
- Release all new items thru the release phases spelled out in the Phase Release Standard unless an executive management written waiver is obtained.
- Assure that modeling and testing that is planned and taking place is appropriate for that phase.
- Assure that the product specification meets the customer's requirements and is released before the product development starts and is continually updated as affected.
- To designate certain team meetings as benchmark/baseline meetings wherein the management (or customer?) is given a presentation as to status, cost, problems, actions, etc.
- Assure that all new items are manufacture-able, test-able, inspect-able, service-able, etc.

- Assure that a first acceptable supplier is designated in the release form for each new purchased item.
- Agree on one structure that satisfies the BOM standard.
- Assure that the team does not try to replace the design engineer's responsibility for the design.

Procedure

- None Required (also see Teams for All CM Processes Standard).

Primary Responsibility

Engineering services manager will keep this standard updated.

Authorization

____

Authorization VP Engineering

20

Phase Release

Purpose

- To specify the company phases of product release. To chart all the critical aspects of those phases. To determine critical phases and elements for the whole company as opposed to any part of the company. To specify the process of Release in a flow diagram.

Applicability

- Applies to all new product and spin-off products to be processed by this CM Group.

Policy/Practice

- The company has analyzed the alternatives and chosen to normally use some or all of the following phases (choose from this list or name your own).
    - Concept Phase
    - Quotation / Contract Phase
    - Design and Development Phase
    - Pilot / Pre-production / Qualification Phase
    - Production Phase
    - In Service (Out of Production) Phase
    - Obsolete Phase
- Phases should be established at the company level — resulting in fewer phases not more phases. The Engineering Design Organization can establish sub-phases and standards as they wish. This standard is for a typical commercial product company.
- The company phases are charted as follows: (add/revise as required - Quotation Phase, etc.)

Phase = Item	**Develop**	**Pilot**	**Production**
Name of Units	Prototype	Pre-Pro	Production
Number of Units to be Built	3 to 6	20 to 30	Per Schedule
Build by	Engr	Pilot Mfg	Manufacturing
Serialized	No	under 100	over 100
Testing	Engr Lab	QA Reliability	Ops Prod Test
Ship to Customers	No	After Upgrade	Yes
Location of Master	Engr	DC	DC
Signatures on Doc	None	Engr	Engr & ME
Revision Level	Date	Numeric	Alpha
MRP Code	D	T	P
ECO to Release	NA	Yes	Yes
Change Control	Engineer	Informal ECO	Formal ECO
ECO Signatures	NA	Engr & ME	Add Field & PC
Changes Master	Engr	DC	DC
Signatures on Changed Master	Engr	DC	DC
ECO Distribution	NA	A list	B list
Interchangeability/PN chg rules	No	Yes	Yes
New Docs Distribution	Engr	Pull	Pull
Revised Docs Distribution	Engr	Pull	Pull

- Costs for building prototype units will be charged to the engineering project. Costs to build pilot units will be charged to pilot manufacturing accounts and costs for building production units to production accounts.
- The vertical separation between phases is often referred to as a "Baseline."
  1. Between prior phase & Design & Development = **Preliminary Design Review**
  2. Between Design & Development and Pilot = **Critical Design Review**
  3. Between Pilot and Production = **Qualification Test Review**
- Appropriate team meetings might be designated as "baseline" reviews.
- Parts, assemblies, design documents and products can only be released (cross a baseline) by a Change Form checked "release for __________ phase."

- After a product is released, if new item(s) are required for a design change, the ECO making the change may also release the new item(s).
- Purchase orders, fabrication orders, or production orders may not be executed for a given phase unless the documentation revision level and the ERP status code match that phase or a "higher" phase.
  Example of use: Don't buy production quantities with a drawing still in the pilot phase.
  Example of use: Buying an item for pilot build may be done if the item is coded Pilot or Production.

Procedure

- See the Release Flow Diagram attached.

Primary Responsibility

- The VP of engineering may take exception to this standard on any particular product by letter/email to the President/GM (copying all key management) outlining the specific exception(s) with reasons given.
- The engineering services director is responsible for keeping this standard current.

Authorization

____

Authorization President/GM

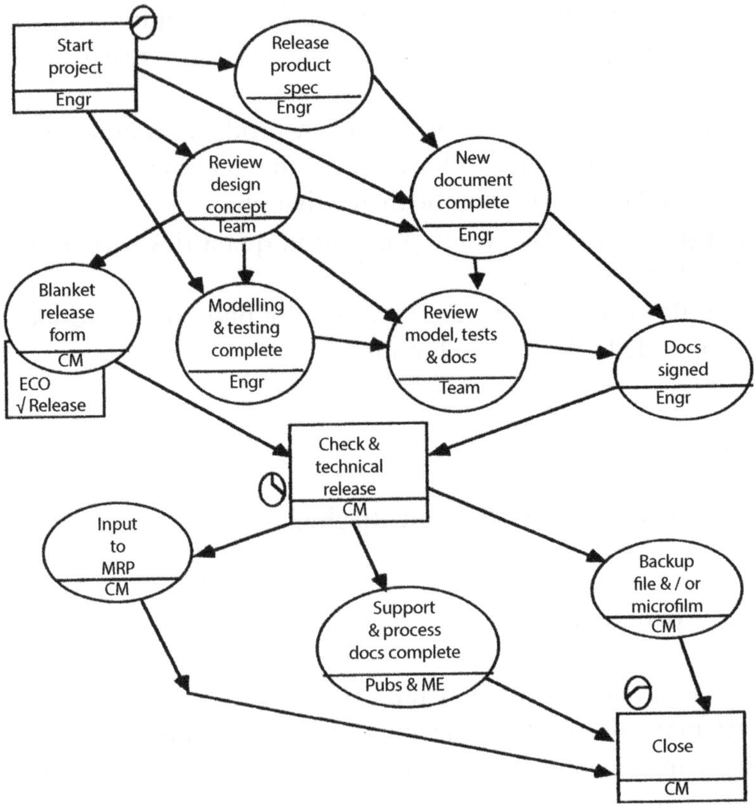
Start
project
Engr
Release
product
spec
Engr
Review
design
concept
Team
New
document
complete
Engr
Blanket
release
form
CM
ECO
√ Release
Modelling
& testing
complete
Engr
Review
model, tests
& docs
Team
Docs
signed
Engr
Check &
technical
release
CM
Input
to
MRP
CM
Support
& process
docs complete
Pubs & ME
Backup
file & / or
microfilm
CM
Close
CM

Part IV
CHANGE REQUEST PROCESS

Definition:

The process which must be followed in order to request action from Design Engineering with regard to the product design and/or documentation.

21

Change Request Policy

Purpose

- To assure that any suspected design problem, design document problem, design suggestion or cost reduction recognized anywhere in the division/company/group is brought as quickly as possible to the attention of the management and the cognizant engineer.
- To assure that Design Engineering responds to the requester in a reasonable period of time as to whether or not engineering will take ownership of the problem or suggestion.
- To reject as many requests as practical since we do not have unlimited engineering manpower.
- To specify the Request for Change Process in a flow diagram.

Applicability

- Applies to all product "problems" for all design cognizant engineers related to this CM function.
- Is not intended to apply to manufacturing process problems, marketing literature problems, or service or manual problems.

Policy / Practice

- Any employee may originate a change request whenever they have a suspected problem, cost reduction or idea concerning the product design or the design documentation.
- The employee with online access may originate a Change Request online. To obtain a description of what must be filled in a form box, place the cursor on the box in question and depress the F1 key or go to the menu "Request Process – Completion Instructions."
- Any employee who is not online may obtain a hard copy form and form instructions (ideally on the back of the form) from the CM function for origination of a Change Request. The completed form should be given directly to CM.

- No signatures, other than the originator, are required in order to submit a Change Request.
- If the employee has obtained the concurrence of the cognizant engineer and is capable of completing the Change Order with the required mark-ups, etc., they should not fill out a Change Request but rather go directly to the Change Order.
- If accepted, the Change Request may later be rejected by cancellation of the Change Order. If this should occur, the CM function will notify the requester and give the reason for cancellation.
- A Management Request Team will include a high-level VP or director from Engineering, Operations and Supply Chain in order to reject as many requests as practical (see Team in Request Standard).
- All requests will be responded to (as to whether or not Engineering takes ownership of the problem) within three workdays on average. Some may take less time and some more time. If the requester wishes to know the status, they may contact CM (Status Report should also be online).
- If your request has been rejected and you are not satisfied with the reasons given, contact CM to state your case. CM will investigate and respond. If you still aren't satisfied, see your manager.
- CM, the Cognizant Engineer and the Request Management Team will respond (take ownership of the problem or not) to requests in a timely manner. They will determine which requests will be submitted to the cognizant engineer according these standards.
- The Cognizant Engineer must use the team in the request process as required by the standard (see Team in Request Standard).
- Document Control will track all requests and sort out non-design issues. Non-design issues shall be forwarded to the responsible function and the requester notified that Design Engineering is not responsible and who is.

Procedure

- See the Request Process Flow Diagram attached to this standard.

Primary Responsibility

- The CM function will track all Change Requests and report to the senior management as well as other affected management as to the average process time, volume and requests in process in aged sequence — oldest first — titled Request Status Report.

- The CM function will be responsible for keeping this standard current.
- The CM manager may take exception to this standard by noting in the change request form why that was done.

Authorization

Authorization	VP Engineering

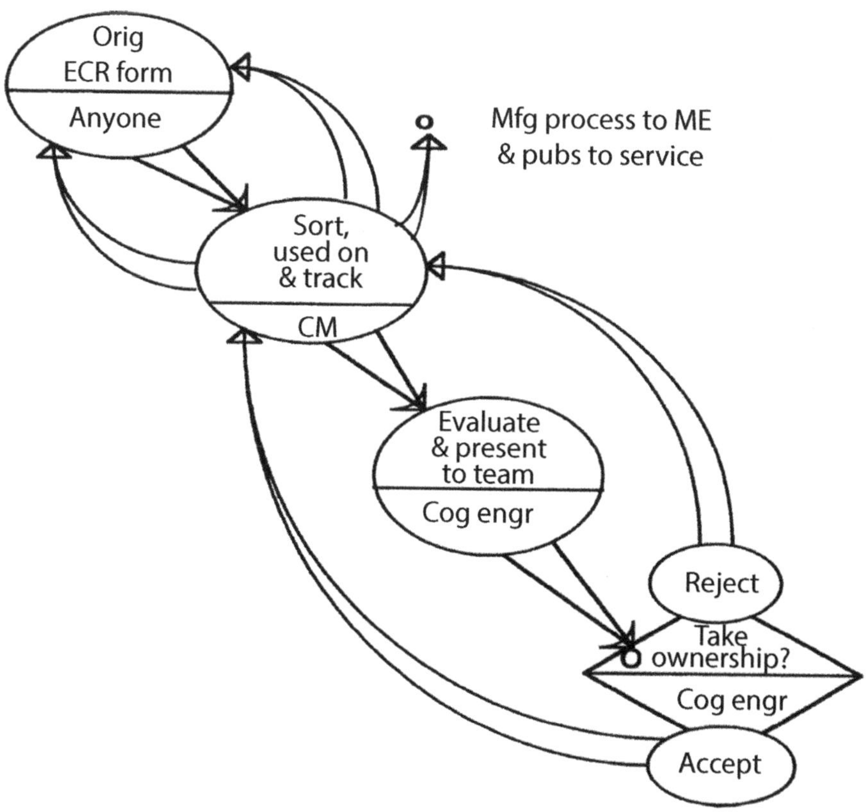

22
Team in Request

Purpose

- To outline the use of a management cross-functional team in the Change Request Process.
- To reject problems and ideas which the team believes: 1. Engineering does not have the manpower to work on it in the next four months, 2. Lack payback according to the company payback policy, 3. Are improvements to products for which the market doesn't demand improvement. (The company should develop a list of products to be/not to be improved.)

Applicability

- To all requests being processed by this CM function.

Policy/Practice

- The Request Team shall consist of a VP of Design Engineering, Operations, and Supply Chain. In case of inability to attend, each VP shall appoint a director as an alternate.
- CM may reject frivolous requests and other requests designated by the team without team review.
- CM may bypass the team on document only changes, interchangeable dimensional changes, changes to meet spec during the Pilot Phase and other categories designated by the team.
- CM and/or the cognizant engineer will present the applicable requests to the team. CM will chair the team meeting and keep a Request Report online, updated and reissued the same day as the team meeting.
- The Request Team meeting will be held frequently and briefly — at least three times a week.
- The team shall reject cost reductions/labor savings that do not meet the cost payback polity as estimated by the team. The team may ask for <u>calculation</u> of the cost of the requested change (what if change is made?) based on tentative solution, tentative effectivity,

tentative disposition of old design parts, and tentative effect on the field (see Costing Design Changes Standard).

- After a request is accepted by the Request Team, the request must be presented to the next Change Team meeting. If the Change Team believes that the request should have been rejected, the cognizant engineer/CM will return the request to the Request Team via CM with the reasoning involved.
- After a change is accepted or rejected CM will notify the Requester.
- The management Request Team may ask the Change Team for their input regarding; members experience on similar problems, manufacture-ability, test-ability, inspect-ability, and service-ability, etc.

Procedure

- None required.

Primary Responsibility

- CM is responsible for keeping this standard current.

Authorization

____

Authorization VP Engineering

23
Request Form

Purpose

- To specify the proper form and form instruction to be used for requesting action from Design Engineering on any product design problem, design suggestion or design document issue.

Applicability

- All requests to be processed by this CM function.

Policy/Practice

- Only the form attached (pg 2) shall be used to request design action. No other form shall be used to request changes from the Design Engineering function. Letters or emails from customers or suppliers are exceptions to this rule — they should be forwarded to the product project engineer.
- This form shall not be used to request changes/improvement to the production process (contact the Manufacturing Engineer).
- This form shall not be used to request changes to quality documentation (contact the Quality Engineering function.)
- This form shall not be used to request changes to the product publications — spares catalog, maintenance, service, etc. (contact the Publications Department)
- The form instruction (pg 3) for completing the Change Request Form may be accessed online by placing the curser over the box to be completed. The form instruction is printed on the back of the hard-copy forms.

Procedure

- See the flow diagram attached to the Change Request Policy Standard.

Primary Responsibility

- The CM Department will be responsible for keeping this standard up to date.

Authorization

Authorization VP Engineering

Engineering Change Request	EC3 Corp	ECR # 1	
Description of Problem/Reason for Request Urgency 2 3			
Problem is with: Product Process Pubs Other 4			
Part Numbers/Document Numbers/Product Numbers (As Known) 5			
Proposed Solution (If you know one) (Attach Mark Ups if possible) 6			
Requesters Name 7	Phone # 8	Page of Pgs 1 9	Date 10
Cognizant Engineer 11	EDC Techs Name 12	Date to Engr 13	
Problem Accepted Rejected Next Product by Cog Engr 14			
Reason for Acceptance/Rejection 15			
Acceptance condition(s) 16			
Title of ECR 17	Cog Egnr Sign 18	Date 19	

Form Date: If you have talked to the responsible engineer and can fully delineate the change, do not fill out the ECR, go directly to the ECO.

Engineering Change Request - Form Instruction

Note: If you have talked to and have the agreement of the cognizant engineer and can fully document the change, do not use the ECR — go directly to the ECO.

Requester	1	Leave Blank (Doc Control will send a copy to you that will be completed thru box #12 including the ECR number).
	2	Describe your problem, suggestion or challenge.
	3	Indicate if the problem, etc., is with the product, the manufacturing process, the publications or other (option).
	4	Enter those part numbers, etc., that you believe to be associated with the problem if you know them — optional.
	5	Describe the solution to the problem if known — optional.
	6	Neatly hand letter your name.
	7	Enter your phone number.
	8	Enter the total number of pages in the set and number all pages accordingly. If no attachments, enter "1".
	9	Enter the current date Month - day - year. Forward the form to Document Control.
Doc Control	10	Sort out process changes and pubs changes and send to those departments. Consult the Cognizant Engineer List for the responsible design engineer, manufacturing engineer or publications person. Enter that name here — hand-lettered neatly. Note that only requests for change in the design will be followed-up by Doc Control.
	11	Enter your name here — hand letter neatly. Copy the form twice.
	12	Enter the date you forward one copy to the Cognizant Engineer and another copy to the Requester. File the master copy for follow-up and reporting.
Cog Engineer	13	Analyze the problem/challenge. Call or see the requester as necessary. Consult the other change request standards and present the problem to the team as/if required and ask CM to respond to the requester accordingly — accept or reject.
	14	Give reasons for acceptance or rejection.
	15	State any conditions on acceptance. Example: "Accepted subject to finding a solution to the problem that isn't worse than the problem."

	16	Give a "name" to accepted problems. Neatly hand letter — use as few words as are practical. (This is the name Doc Control will use when reporting on the ECR status.)
	17	Sign the ECR and make a copy.
	18	Enter the date signed. Return a copy of the ECR to Doc Control.
Doc Control	18	Log the date, make a copy and send it to the Requester whether accepted or rejected. Report as required by the change request policy.

Part V
CHANGE CONTROL PROCESS

The process for changing the product design and/or the associated documentation.

24

Change Control Policy

Purpose

- To define the policy and practices required to process product design and documentation changes in a fast, accurate and well-understood fashion.
- To specify the Change process work flow in flow diagrams for the three major phases of the change process (see attachments).

Applicability

- Applies to all design changes from cognizant engineers related to this CM function after release of an item to Pilot Production or Production.
- Not intended to apply to manufacturing process changes, marketing literature changes, or service/manual changes.
- Does not apply when the product is in Design and Development Phase.

Policy/Practice

- The change form must be used to release (see Release Standards) a new design document or item or to change a design document or an item, or to both release and change when the change requires a new item(s).
- The change form shall be the only way to make changes to a design document or an item. Other forms, such as the Deviation, Waiver, Change Request, etc., shall not be used to make design changes. If redlining on the production floor is allowed, it must only be done by the Line-Down Change Standard.
- Each change form is identified with a unique number assigned by CM normally after the Check Point is successfully passed. (Gives assurance to all affected that the change will be promptly released.)

- The cross-functional team will be used to review certain requests and changes. See the applicable Team Standards.
- Certain signatures are required per the Signatures Standard.
- Those changes identified as "cost reductions" must have the cost calculated per the Costing Design Changes Standard.
- After passing the "Design Complete" event, no changes to the technical aspects of the change, change form or attachments will be allowed. After that event, the change will be processed as is or canceled. Exceptions are allowed only per the "Primary Responsibility" of this standard.
- All non-interchangeable changes must be modeled and tested on a production product. The VP of engineering must approve any exception to this rule.
- One problem fix/improvement shall be delineated on one form with all necessary new or marked-up documents included in that change. No more than one problem/improvement will be included in one change. Multiple change forms for one problem/improvement will not be allowed. Exceptions to these rules are as follows:
  1. Software or firmware changes (not dependent upon any hardware changes) may be combined into logical groupings and tested and "released" as a group.
  2. Several "Document Only" changes to the same document may be combined into one change. Such changes may be "queued" (held for the next document-only change) only by the cognizant engineer.
  3. Several interchangeable changes to the same item providing they can be made effective together. Such changes may be "queued" (held for the next change) only by the cognizant engineer.
  4. Short-term fix and long-term fix may be on two change forms with the short-term fix done now and the long-term fix done later.
  5. This policy allows for a fix to be done across the product line or across a document type by a "blanket" change form. Blankets will normally be done in hard copy offline so that they do not affect the thru-put time measurement or the aged reports.
- Use of marked prints is required in all cases except where the change can be fully detailed by "From-To" delineation in the space for "Description of Change" in hard copy box (see Mark Up of Design Documents Standard).
- Use of CAD "overlays" or "redlines" or word processing "redlining" techniques is encouraged instead of hand mark up.

- The Change Request number or other reference numbers should be shown on the change form, but such documents need not be included in the change package.
- If unique rework (other than "Rework to Print") configurations are allowed, they should be included in the change package before the "Design Complete" event.
- New part, assembly or related documents required as part of the change must be included in the change package.
- All affected documents must be included in the change package — revised pages only.
- The impacted functions will be determined by and per the Impacted/Affected by a Change Standard.
- Masters of change documents (online or hard copy) will be controlled, and kept by the cognizant engineer prior to "Design Complete" and by CM afterward. They will be available for review by any interested party. After release to Manufacturing they will be filed by the change form number in/by CM.
- The source of the used on / where used will be the ERP BOM database.
- Add any policy statements required as a result of ERP or PLM system design here.
- The change system phases are defined as follows:
  1. Design (Redesign) Phase – From "Start" to "Design Complete".
  2. CM Phase – From "Design Complete" to "Release to Manufacture" — the latter of ERP Updated or Master Documents Updated.
  3. Implementation Phase – From "Release to Close".
- The company goal for the average processing time through the CM phase is five workdays.
- Time limits for performing certain functions including signatures are: (add your own time limit standard)
- The originator of the request, change or any other party that believes that the change must be expedited, may walk to/email/telephone those involved in the process to accomplish this. CM will not hand-carry or expedite any given change but will expedite the whole of changes.
- Problems encountered during the processing should be brought to the attention of the responsible CM technician.

Procedure

- See the flow diagrams attached.

Primary Responsibility

- The CM function is responsible for maintenance of this standard in a current condition.
- The CM manager may take exception to this standard on any given change by noting the reason(s) for exception on the applicable change form.
- CM will maintain the workflow diagrams.

Authorization

Authorization	VP Engineering

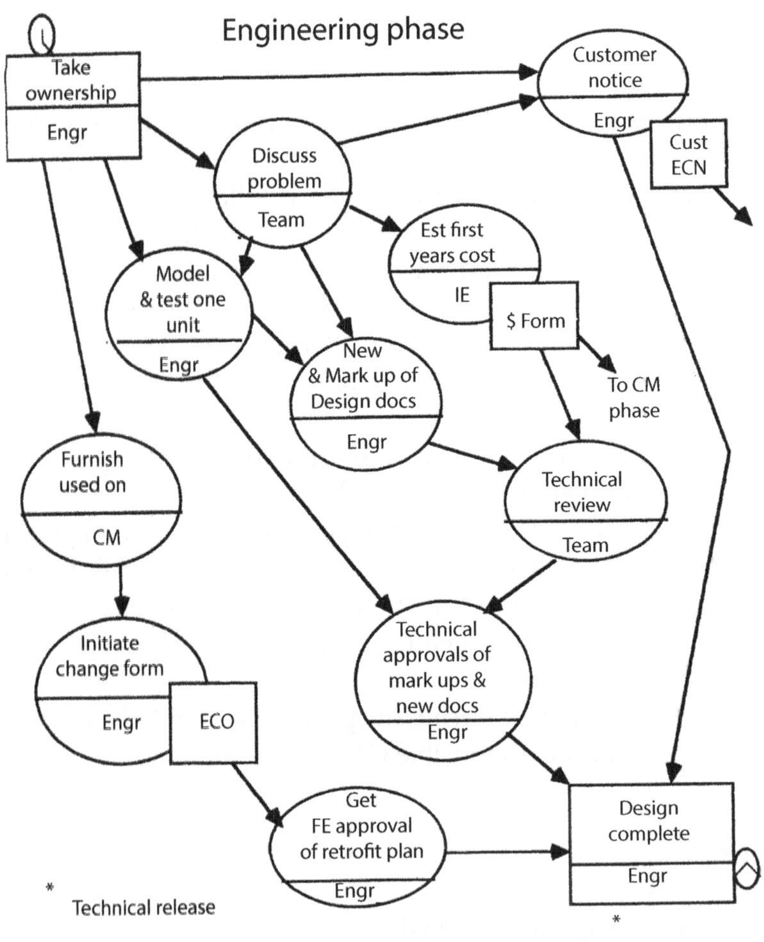

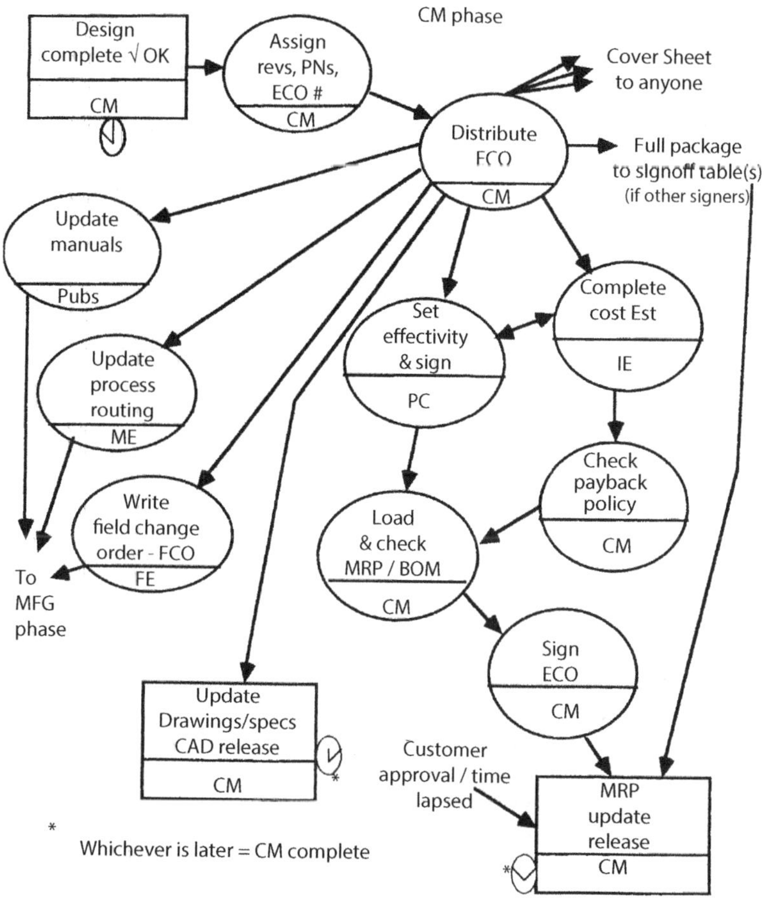
CM phase
Design complete √ OK
CM
Assign revs, PNs, ECO #
CM
Cover Sheet to anyone
Distribute FCO
CM
Full package to signoff table(s) (if other signers)
Update manuals
Pubs
Complete cost Est
IE
Set effectivity & sign
PC
Update process routing
ME
Check payback policy
CM
Write field change order - FCO
FE
Load & check MRP / BOM
CM
To MFG phase
Sign ECO
CM
Update Drawings/specs CAD release
CM
Customer approval / time lapsed
MRP update release
CM
*
Whichever is later = CM complete

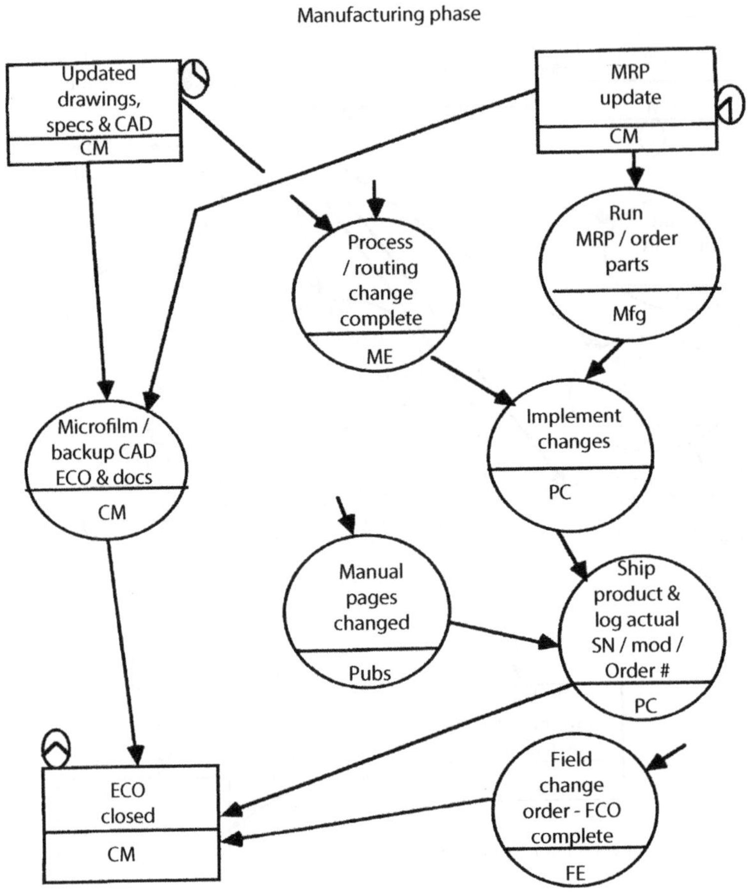
Manufacturing phase
Updated drawings, specs & CAD
CM
MRP update
CM
Run MRP / order parts
Mfg
Process / routing change complete
ME
Microfilm / backup CAD ECO & docs
CM
Implement changes
PC
Manual pages changed
Pubs
Ship product & log actual SN / mod / Order #
PC
ECO closed
CM
Field change order - FCO complete
FE

25
Team in Change

Purpose

- To specify the makeup of the Design Change Team.
- To outline how the team is to be used in the change process.

Applicability

- All changes to be processed by this CM function.

Policy/Practice

- The team shall consist of one representative from each function which might be affected by a change. Each representative will have an alternative assigned by their management.
- CM will chair the team meeting and produce a change order Action Items List updated the same day.
- All changes (except Doc Only) will be brought to and discussed briefly by the team as soon as the Change Request has been accepted by the Request Team or by Design Engineering management. If there is no Change Request, the cognizant engineer must bring the issue to a team meeting for discussion as soon as he/she is aware of the issue (may or may not have started a change order).
- All change redline documents (except Doc Only) must be brought to at least one Change Team meeting. All change redline documents and the change document must be signed per the Signatures Standard either in the meeting or obtained by the cognizant engineer outside of the meeting.
- All changes identified as "cost reductions" or "non-interchangeable" must be reviewed by the team for required/possible calculation of the cost.
- The team with the concurrence of the cognizant design engineer may cost any change, hold any change for costing, or determine to proceed without the cost finalized. If required by the team, CM will assure that the cost/payback is finalized prior to Release to Mfg (see Costing Design Change Standard).

- The cognizant design engineers may bring a change to the team as often as they wish prior to "Design Complete." After Design Complete, CM may bring the change back to the team as needed.
- Any change with an open action item will be discussed by the team when the action date is reached.
- The team is not responsible for product design, the cognizant engineer is (see Teams for All CM Processes Standard).
- No team discussion/meeting is required on a change after "Design Complete."

Procedure

- None required.

Primary Responsibility

- The CM manager shall keep this standard current.
- CM technicians can take exception to these requirements providing the reason is noted on the change form.

Authorization

Authorization	VP Engineering

26

Change Form

Purpose

- To specify the only/proper form and the instructions to be used to make product design or document changes.

Applicability

- All product design or document changes to be processed by this CM function.

Policy/Practice

- Only the form attached (see pg 2) shall be used to make product design or document changes.
- The change form shall not be used to make changes to process, quality or sales documentation.
- The change form shall not be used to make changes to the product publications (spare parts lists, service manuals, etc.).
- The proper form instruction is found on page 3 (should also be on the back of hard copy forms) and may be seen online by placing the curser over the form box.
- No other form shall be used to make design or design documentation changes.
- This form shall be used to release documents individually or in groups (in lead time).
- If new items are required as part of a design change, they may be released with this change form.
- In recognition that "paperless" is often impractical, "less paper" should be our goal. Therefore, the design of online forms, when printed, will normally be one 8 ½ × 11 size, rather than the "run-on" design used by most IT departments and systems.
- If your system allows, redesign the online form to accommodate the "less-paper" reality.

Procedure

- None Required.

Primary Responsibility

- CM will be responsible for keeping this standard and the form design up to date.

Authorization

____

Authorization VP Engineering

Engineering Change Order	EC 3 Corp	CHANGE ☐ RELEASE ☐ 4	ECO # 1

Class
☐ Doc - Document Only
☐ I - Interchangeable
☐ NI - Noninterchangeable 6

Justification for change: 5

ECR # 2

Type: Meet Prod Specs ☐ Reduce Costs ☐ Exceed Prod Specs ☐ 7

Effective Point:	Next Order	Next P O	At Vendor	Whse Issue	In Asm	In Test	Burn In	System Test	In FGS	In Field 8

Effectivity:
Plan: Replan: Actual: 29

Applications Affected: 9

Background / Facts: 10

Customer / Agency Review or Approval Plan: 11

Retrofit yes ☐ no ☐ If yes state plan: 27

Rev Reason: ¤ C ¤ A ¤ R 36

Impacts / Affects:

		Yes	No
Regulators	Reg	☐	☐
SW/FW	SE	☐	☐
Tools/Fixtures	ME	☐	☐
Test Equip	TE	☐	☐
Assy Process	IE	☐	☐
Supplier(s)	SC	☐	☐
Packaging	IE	☐	☐
Publications	FS	☐	☐
Product Spec	PE	☐	☐
		12	

Detailed Description of Change: (From To Including drawing zones) (If more space is needed, attach mark ups) 13

Modeled & Tested: SN: Date: Report # : Test By: 14

Documents Affected					Old PN Disposition				
Old PN	Rev	New PN	Rev	Noun Name	Use	Scrap	RTV	Rwk able	Mfg Rwk
15	16	17	18	19	22	23	24	25	26

Start Date 3	Design Compl. 21	MRP Updated 30	Masters Updated 32	Closed Date 33	
Cog Engr Sign 20	Mfg Eff Sign 29	Field Sign 28	CM / Mgmt Sign 31	Rev Date 35	Pg of Pgs 34

Engineering Change Order – Form Instruction

Note: Anyone may fill out an ECO for the cognizant engineer if they have talked to the engineer and have both agreed to do an ECO.

Note: Steps may not be performed in the sequence shown (see the Flow Diagram).

Cog Engineer	1	Leave blank. CM will assign an ECO number after receipt/check/design complete.
	2	If there was an ECR that preceded the ECO, enter its number here.
	3	Enter the date that you first realized that the problem being fixed was significant enough to warrant a design change.
	4	Check "Release" if document(s) are being released (see Product and Document Release Chapter). Check "Change" if document(s) are being changed. Check both if the change includes release of a new document(s). If the ECO is a "pure release" then boxes 2, 3, 6, 7, 10, 11, 13, 14, 15, 16, 22, 23, 24, 25, 26, 27, 28, 29, 30, 32 & 36 may be left blank.
	5	Justify, in the space provided, why the change is necessary.
	6	Check the correct class. One check allowed (see Interchangeability Standard). If the change is a "Doc Only" then boxes 7, 8, 12, 13, 17, 18, 22, 23, 24, 25, 26, 27, 28, 29, 32, 33 & 36 may be left blank. If the change is "interchangeable" then boxes 14, 27 & 28 may be left blank.
	7	Check the type of change. More than one check is acceptable.
	9	Enter the product or model number(s) that are to be changed. The ERP system is the only acceptable source for "used-on."
	10	Give facts/background about the number of units that have failed, field population, customers affected, etc. Explain safety issues. Time/costs to be saved. Attach relevant documentation.
	11	Is customer/agency approval required? Review? Not applicable? If approval is required, has it been obtained? If not obtained, should the

change proceed or be held for approval? If customer review is required, has a description of the change and other required data been sent.

Cog Engr Continue 12 Review the impacts list and check each line, either yes or no.

13 a) Precisely describe the change in "from-to" detail. Include drawing zone for each change. Example: In zone C - 12 change length dimension from 14.000 in to 14.100 in. Change its tolerance from + or - .006 to .003.

If space provided is not adequate, attach neatly marked-up prints or specs. Mark up of a released Engineering Parts List is allowed and encouraged (see Markup of Design Documents Standard).

Any combination of "From-To" or marked prints is allowed.

The cognizant engineer must obtain the Manufacturing Engineers signature on the "From-To" in block #13 and/or on the marked prints attached to the ECO.

34 Complete according to the total pages in the set. Number each page similarly.

14 For those changes which must be physically modeled and tested (see ECO policy), indicate the SN modeled, the date it was tested, the report number / notebook page(s) / file number that the results are shown in and indicate who performed the test.

15 List each document affected by the change whether it is being revised or the part number is changing. If a marked parts list is attached, enter "Marked PL attach". (Document Control will either enter boxes 15, 16, 17, 18 and 19; or will attach an "Old PN Disposition" to the parts list; or allow Production Control to add the part disposition to the PL mark up.)

16 Show the current revision level of each part number/document affected by the change.

17 If parts/documents are being released, show their numbers here. If a part number is changed, show the new part number opposite the old part number.

	18	Show the revision level of each new part number (rev numeric for pilot release, rev alpha for production release). ERP system is official source.
	19	Enter only the noun name of the item being revised or released.
	25	Indicate if the old PN can physically be reworked into the new PN.
	27	The field service engineer and the design engineer must agree on the retrofit plan (on repair, on failure, at next maintenance or recall, etc.).
	28	Obtain the field service engineer's signature on the form if retrofit is directed.
	20	Sign the form.
Cog Engr/Doc Control		
	21	The date you give the form to Document Control. (Doc Control is authorized to update this box to the date received.)
Document Control	1	Check the form and attachments according to the Change Design Complete Standard and all other applicable standards.) If OK, assign ECO #. If not OK, return to the cognizant engineer with specific deficiency notations. When acceptable, distribute the complete package online to all potentially impacted/affected (whether checked yes or no).
All Impacted	12	All on the list, whether the engineer checked yes or no, must take ownership of the check or contact Doc Control to correct the check. One day will be allowed by standard. Also notify Production Control of the lead time in work days to perform the work required — 2 days will be allowed by standard.
Production Control	22	The quantity (or check – you decide) of old design item to be "used up/as is."
	23	The quantity (or check) of the old design part to be scrapped.
	24	The quantity (or check) of the old design part to be "Returned to Vendor." Check all deleted parts for other where used / usage /stock status. Return those that are excess/obsolete now since later they will be less likely to be returnable.
	26	The quantity (or check) of the old design part to be actually reworked (allowing that manufacturing may decide whether or not to rework).

	8	Circle the point in the manufacturing process at which the change will be made effective.
	29	Plan the effectivity date for the change. (Also responsible for following up on that plan, changing the plan as required by entering a "replan" date, and capturing the actual date on which the change was incorporated into the product.) Sign the ECO after all the responsible boxes have been completed.
Document Control	30	Input all the required changes and additions to the ERP Item Master file and BOM file for design data elements only — including the effective date. (Screens _____ in the ERP) Enter the date complete (output/screen must be checked by a different person than did the input).
	31	Sign the ECO and distribute the cover sheet to all affected (check "Yes" in box #12).
	32	Update all the master documents/files affected by the change and check by a different person than did the change incorporation.
Production Control	29	Track the change to actual incorporation into the product and give that Date/SN to Document Control. Update applicable data in the ERP screens. This applies to all who have ownership of an ERP screen.
Document Control	35	Enter the actual effective date in the ERP and into the master ECO box #29. Rev the ECO date in box #35. Redistribute the ECO cover sheet to anyone who needs to know the actual effectivity.
	36	Categorize each new ECO and each revision to an ECO as follows: C = This ECO is a Correction of the design (from-to or mark ups) in an earlier ECO. A = An Administrative change to effectivity plan, impact yes- no checks, etc., after box #31 distribution. R = A Redesign of the fix (from-to or mark ups) after completion of box #21. Redistribute the cover sheet to all affected – "Yes". File the master package by the ECO #.

27
Interchangeability

Purpose

- To define interchangeability and the resulting actions required.

Applicability

- All changes to be processed by this CM function.

Policy/Practice

- Interchangeable – Two or more items are considered interchangeable if, in all applications, they are:
  1. Of an acceptable **form** (appearance) to fulfill all esthetic requirements per the **product specifications.**
  2. Of a proper **fit** (physical) to assemble with other mating items per the **drawing dimensions and tolerances.**
  3. Of a proper **function** to meet **product specifications**, including performance, safety and reliability requirements.
- Items meeting these criteria are completely interchangeable one for the other (both ways) with no special adjustments, modifications, or alterations to the item or related items.
- Items that meet some but not all of the above criteria are not completely interchangeable and are therefore considered non-interchangeable.
- Compatible is defined to mean that the old is not interchangeable in the new but the new is interchangeable in the old.
- Whether part numbers or revision levels change is determined by use of the Part Number & Rev Changes Logic Diagram (see Part Number & Revision Level Changes Standard).
- Non-interchangeable changes do not <u>always</u> require scrapping or reworking – disposition of old design parts is a separable decision (see Disposition of Old Design Parts Standard).
- Only those non-interchangeable changes designated to "retrofit" will be installed into returned goods or field units. "Update everything to the latest revision level" is very expensive and will not be done.

- If the product specifications are not clear when processing a change, they should be revised to clarify the interchangeability. If the product specifications are to be revised, it should be in the same change that raised the issue. If the cognizant engineer chooses not to revise the specification to clarify the issue, that change must be considered interchangeable.

Procedure

- None required.

Primary Responsibility

- The CM manager is responsible for keeping this standard current.

Authorization

____

Authorization VP Engineering

28

Part Number & Revision Level Changes

Purpose

- To establish rules as to when to change part numbers and when to change revision levels.
- Changing part numbers when appropriate avoids making a rule for retrofit that says, "all units will be retrofit to the latest revision level."

Applicability

- All changes processed by this CM function.

Policy/Practice

- The following logic diagram will be used to determine Part Number and Revision Level change:

PART NUMBER CHANGE LOGIC

FRU = Field Replaceable Unit / Service Assembly / Spared Assembly

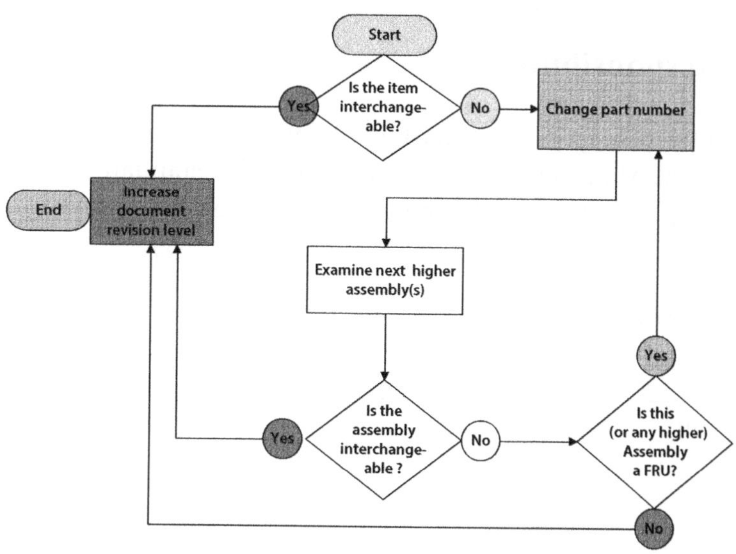

- Consider all applications of all the items changing.
- Assemblies that are not field replaceable units (FRU) need not change Part Number even though they are not interchangeable (per the logic diagram).
- "Part Number change" will mean a change of only the tab portion of the Part Number.
- Top assembly (Product) will not change Part Number unless Marketing wishes to sell both versions.
- Part Numbers need not change on non-interchangeable change if no parts have been made or if Production Control indicates that all parts made will conform (Pilot Phase parts included).
- Phantom assemblies do not need to change Part Number on non-interchangeable change.
- A parts list change may be interchangeable at the assembly level and, if so, it is a revision level change to the assembly drawing/parts list.
- Parts don't have revision levels, drawings do. Parts are made from drawings at a particular revision level. If your ERP system demands a rev level change, fool it with a 99 or XX or whatever.
- Revision level is not a part of the Part Number.
- A blind person working in the stockroom with a pick list part numbered in Braille and with the bins part numbered in Braille, must be able to find interchangeable parts. (Most ERP systems don't issue parts by revision level — thus the ERP is the "blind person.")

Procedure

- None required.

Primary Responsibility

- CM must keep this standard up to date.
- The CM Manager can take exception to this standard by stating in the applicable change form why it was done.

Authorization

Authorization	VP Operations

29

Change Classification

Purpose

- To define the acceptable change classes.
- To allow exceptions to or skip steps in the change or other processes based upon class.
- To help identify changes wherein non-interchangeable Part Number changes should be considered.

Applicability

- All change forms to be processed by this CM function.

Policy/Practice

- General Class
  1. Doc Only – see definitions.
  2. Interchangeable – see applicable standard.
  3. Non-interchangeable – see applicable standard.
- Modification to the product which is necessary to make the product:
  1. Meet the product specifications or safety standards.
  2. At a reduced manufacturing or maintenance cost.
  3. Exceed its specification.
- Modification to the product design might be a combination of 1, 2 or 3 above.
- Any change identified as required to meet product specifications or safety specifications must also be considered non-interchangeable and Part Number change should be considered (see Interchangeability Standard and the Part Number & Revision Level Changes Standard).
- Other classes allowed as defined by the CM manager might include (Note: the fewer classes the better):
  1. Urgency (providing it takes no more than 10 man-minutes to determine).
  2. Cost effect on the product.
  3. Etc.

Procedure

- None required.

Primary Responsibility

- The Doc Control Department shall have responsibility to keep this standard up to date.

Authorization

____

Authorization CM Manager

30

Mark Up of Design Documents

Purpose

- To encourage the use of the mark ups for communication of the difference between the old and new designs. This saves time (over the "from-to" method) on the part of all users of the change package. It improves the clarity of design change over the "from-to" or other methods.
- To specify what an acceptable mark up is.
- Knowing precisely the differences between two revision levels of a document is extremely important to: suppliers, fabricators, various manufacturing and field service functions as well as future design problem troubleshooting. Carefully documenting the differences in mark ups is paid back many times by these people who can avoid the need to compare two revision level documents to tell the differences or to interpret from-to descriptions
- Use of the mark-up technique is encouraged since it affords the fewest possibilities for transposition errors. The master mark up (in red) will be used by the CM Drafts draftsperson/ CAD Designer designer to revise the master document.

Applicability

- All Change Orders processed by this CM function.

Policy/Practice

- A print or CAD copy of the latest revision level document must be used for mark up.
- CAD "overlay" or "redline" ability serves as a very desirable mark-up method — as opposed to hand mark up (If you own it, use it, if not, buy it.)
- A very high quality latest revision print can be obtained from the PLM System (or Drawing Center on a "while you wait" basis — ask for a "print for mark up").

- After the "design complete" event, the revised master document will not be returned to the responsible engineer for signing or initialing. The change to the master documents will be made by CM in exact accord with the mark up. (Encourages engineers to "do it right" the first time.)
- Mark ups must be clearly and neatly done since they will be used in the Change Order "as is."
- Pictorial, parts lists and text documents should all be marked up as part of a Change Order — by highlighting the portion to be deleted and entering the new portion in script, or use of "track changes," or by any other definitive word processing method.
- ERP parts lists (see Bills of Material Standard) must be double spaced for mark up clarity. If your ERP system doesn't print double-spaced parts lists, it should be programmed to do so.
- Hand mark up of the new condition must be neatly done in red pen or pencil, bold and printed (not cursive). It should be in a different lettering "style" than the original lettering.
- Both the print and the mark-ups must be capable of being readable after three generations of reproduction . . . to silver microfilm, to diazo microfilm, to print. (If you don't use microfilm, substitute correct number of generations and wording.)
- Portions of a document may be "cut and pasted" providing:
  1. Standard drawing sizes are used (multiples of 8½" × 11")
  2. The print is the latest revision level in the portion used.
  3. The title block is included in the cut and paste mark up.
- CM may also cut and paste mark ups in order to reduce the marked-up document size.
- The "old" ("was" or "from") condition shall not be obliterated in part or in total. Underlining the portion to be deleted is the only acceptable method.
- Mark-ups can be changed only at or before signing / design complete.
- The mark-up technique can be used for the "make from" (same as except) creation of a new item/document.
- Encouraging and preferring the use of mark-ups does not preclude the use of "from-to" descriptions. When the change can be completely and precisely described on the Change Order form by "from-to" in the space provided on a hard copy form. If this is the case, do not mark up the drawing — do the "from-to" description. The drawing zone and an exact description of old and new configurations are required in this case.
- A complete set of acceptable examples of mark-ups of all company documents is maintained in CM for reference. (Put them on a wall and walk every new engineer and designer past the wall.)

- Only the affected pages of multiple page sets need be marked up and included in the change document.

Procedure

- None required.

Primary Responsibility

- CM will be responsible for maintaining this standard in a current condition.
- CM technicians have the authority to take exception to this standard on individual Change Orders by explaining in the Change Order Form why exception was required.

Authorization

____

Authorization VP Engineering

31
Effectivity

Purpose

- To define the point in the production process and the date at which a given design change will be implemented.
- The date at which the change is / is planned to be effective.

Applicability

- All changes to design/design documentation by this Document Control function.

Policy/Practice

- The following effectivity points will be shown on the change form and are defined to reflect the point in the process at which the new design will be implemented:
  1. **Next Order** – Next order placed to a supplier if a purchased item or to the fabrication shop if an item is fabricated in house.
  2. **Supplier** – Items on order are to be affected by Buyer action with the Supplier(s). It is expected that the supplier will be changing all items not yet fabricated by the effective date. If the item is fabricated in house, Operations must change all items not fabricated by the date specified.
  3. **Receiving Inspection** – All items received from supplier(s) on or after the effective date must contain the change (the easiest point to verify for purchased items).
  4. **Issue** – All issues from the warehouse on or after the effective date specified. (Most ERP systems use this date for parts list changes.)
  5. **In Assembly** – The items built on or after the date specified.
  6. **In Test** – All items tested on or after the date specified.
  7. **In Burn In** – All items placed into elevated temperature run in / test on or after the date specified.

8. **System Test** – All items put into system test on or after the effective date.
9. **FGS** – Affect all items in Finished Goods Stock. Must have the approval of the Operations VP to specify this point.

- Items built (fabrication or assembly) or tested prior to the effective date need not be changed if recycled for rework prior to shipment.
- All items in the manufacturing pipeline prior to the effective point will be affected by the change.
- The customer returned goods area will not be affected by the effective point/date. That is, if returned goods are to be affected it must be by retrofit notations on the change order.

Procedure

- Not applicable.

Primary Responsibility

- The Production Control Department will be responsible for maintaining this standard.
- Production control planner may take exception to this standard by noting on the change form why exception was taken.

Authorization

Authorization VP Operations

32
Effectivity Management

PURPOSE

- To outline methods and policies for determining the effective date of each change.
- Will be used for planning/recording the effective Serial Number, Order Number, or Lot Number, or Date Code, etc. (edit according to your process).

Applicability

- All Releases and Changes done by the change form for this CM function.

Policy/Practice

- Each change (one problem, one fix, one change order) will also require one effectivity point on the change form (see Effectivity Standard).
- Each change will require one effective date on the change form (one problem, one fix, one effective date) for each product affected.
- If required by a customer, the effectivity of the change will be planned by SN, Order Number, Lot Number, Date Code, etc. (edit accordingly).
- Several interchangeable changes to one part may be made on the same change form providing they can be effective together.
- Upon receipt of the change, each impacted department has the responsibility for notifying Production Control of their lead time to change. Each impacted department must notify Production Control immediately upon recognizing that their lead time is changing. Notification will be done (online or by telephone or in the Team Meeting — pick your method).
- Production Control has the overall responsibility for this management task as well as the responsibility to track all affected material. They will track all the parts and assemblies affected by the change to assure that the change occurs as planned or that the plan is revised (re-planned) as necessary.

- Production Control will notify CM by email as to the original plan, any re-plans and the actual effectivity (date or SN, etc.). (If your system allows, consider PC entry of the plan directly.)
- Document only changes do not have an effective date. NA will be entered on the change form by CM. Several "document only" changes can be made to one document by one change form.
- Parts list changes will normally have an effective point at the issue date from the warehouse, the date being the issue of the new items. (Some ERP systems use the date as the last issue of the old design items — correct as necessary.)
- Customer requirements that are stated in the change form by Engineering will be followed unless Production Control negotiates revised requirements.
- Production Control must balance the cognizant engineer's preferences, lowest cost and material availability in setting the effective date.
- Production Control will plan the initial effective date, re-plan, track the change as necessary and enter the actual effectivity as follows:
  1. **Interchangeable changes with the actual date** the change occurred at the specified point in the production process.
  2. **Non-interchangeable changes with the actual serial number(s)** that contain the change - Example: SNs 682, 685, 686, 689 & up. (If you trace by date code, order number, lot number, etc., modify this statement as required.)
- If you are trying to set an effective date by line item affected (not recommended), at least the following conditions are possible:
  1. You may be trying to put more than "one problem / one fix" into one change (not recommended).
  2. You may be trying to do Production Control detail planning on the change form. If you wish to do this, use page two of the Change Form to do this or do that detail planning offline.
- The materials planning is an extremely dynamic situation and therefore requires Production Control to revisit every open change for each product affected at least weekly to determine if the date(s) need to be re-planned and to notify CM if the planned date(s) changes.

Procedure

- See change flow diagram.

Primary Responsibility

- Production Control is responsible for keeping this standard current.
- Production Control can take exception to this standard with the concurrence of the CM manager and stating in the change form why such exception was taken.

Authorization

Authorization VP Engineering / VP Operations

33
Disposition of Old Design Parts

Purpose

- To define the acceptable methods to be used to disposition old design parts upon making a change.

Applicability

- All changes to be processed by this CM function.

Policy/Practice

- Disposition of the old design parts shall be done in four categories by Production Control:
  1. **Use As Is / Use Up / Deplete** – Implement the new design item use immediately after using up the current inventory and work in progress. Buy/build no more of the old design item.
  2. **Scrap** – Throw away the current inventory. Enter quantity to be scrapped on the change form.
  3. **Return to Supplier** – The current inventory (quantity indicated) is to be returned to the supplier.
  4. **Rework-able/Rework** – See below.
- The cognizant engineer should specify whether or not each old design part is physically rework-able — yes or no.
- Production Control/Manufacturing Engineering should decide whether or not to rework since it may be more expensive to rework than to obtain a revised item or the expense may be a secondary consideration to keeping the production line going.
- Production Control, with Purchasing, should determine whether or not to return the old design parts to the supplier. Although most parts to be returned to the supplier are done via MRB action, changes must be reviewed for the same possibility — especially deleted parts on a parts list change that have no other "where used/used-on".

- Disposition is noted on the change form by entering a "yes" or "no" or quantity of parts under the proper code. (Edit to your choice.)

Procedure

- None required.

Primary Responsibility

- Production Control is responsible for keeping this standard up to date.
- Production Control may take exception to this standard by noting on the change form why this was done.

Authorization

____

Authorization VP Operations

34

Impacted/Affected by a Change

Purpose

- To outline the responsibilities for those who <u>might be</u> impacted by any given change.

Applicability

- All changes to be processed by this CM function.

Policy/Practice

- All functions which can be impacted by a design change shall be listed on the change form with a □ Yes □ No
- All functions will be reviewed by the cognizant engineer when a change is originated and each line checked either yes or no according to the engineer's best estimation as to whether or not they will be affected. (Causes engineers to consider the potential impact of each change.)
- All functions will receive a copy of the change form and must take responsibility for the accuracy of the yes or no check within 24 hours — if the engineer's check is incorrect, correct the yes or no check online (telephone CM if not online — do not use email).
- If the affected functions make no correction during this first workday, the engineer's check will be correct per policy and the affected function now has ownership of the yes/no check.
- If your function is affected (yes), within another/next 24 hours, email Production Control with the lead time to implement, completion date and briefly describe the task(s) required.
- Each affected (yes) function has the responsibility to update Production Control with their lead time and/or the completion date whenever the conditions change.
- Production Control will use the lead time / completion and their materials analysis to determine the effectivity of the change and

to telephone the effectivity to CM / enter online. This must occur within 48 hours.

- Use of the lapsed time (two 24-hour steps) feature is strongly encouraged and must eventually be reached if the total CM time is to reach five workdays.

Procedure

- None required – See change control form and flow standards.

Primary Responsibility

- The CM manager is responsible for keeping this standard up to date.

Authorization

Authorization:	Engr Services Director	Production Control Mgr

35

Product Improvements

Purpose

- To define which products to improve and under what circumstances.
- Also, to define which should not be improved.

Applicability

- Applies to all products released by this CM function.

Policy/Practice

- A list of all company products shall be made and each product will be identified as to whether or not it is to be improved.
- Certain products need to be improved to "meet the market." They will be indentified as Class X.
- Certain products need to be improved only if the improvement is a real cost reduction exceeding $xxx per unit and/or pay back in yy months (see Costing Design Changes Standard). They will be identified as Class Y.
- Some products should not be improved as the profit margin is too thin or the market does not demand improvement. They will be identified as Class Z.

Procedure

- CM manager will keep the list and will ask the executive management to update it at least semiannually. CM will make the list product class available to only those engineers who need to know without reproducing the list and without publication of the list of any kind.
- Marketing VP, engineering VP and the general manager shall review the list semiannually or more often if desired.

Primary Responsibility

- The director of engineering services will be responsible for keeping this standard up to date.

Authorization

____

Authorization Engineering VP

36

Change Design Complete

Purpose

- To specify the major events in the Change Process and to show the Design Complete event in the work flow. To define the requirements for a change to be called "Change Design Complete" — a flag for those impacted to begin work on implementation of the change and that point at which an ECO Number will be assigned.

Applicability

- All design changes to be processed by this CM group.

Policy/Practice

- In order for Doc Control to accept a change as being Design Complete, it must meet the following **Checklist Criteria**:
  - If the change is being made to reduce cost, the cost must have been calculated per that standard. That calculation must be complete as determined by the cognizant engineer and the change team.
  - If the change is being made to meet product or safety specifications, have the specifications been checked to assure that the applicable criteria is in those specifications.
  - If the change is "to exceed specifications," has the VP of engineering signed the change and is the product affected on the "To Be Improved" list. (Create a standard listing all products in three categories: 1. Improve to meet the market or 2. Improvement allowed if it is a real cost reduction. 3. Do not improve.)
  - If the change is non-interchangeable has the change been modeled and tested?
  - As required by the change team standards, has the team reviewed the problem and the fix.
  - Once a change is accepted by CM as "Design Complete," all who need to begin execution of the change in their impacted functions must do so. This point can thus be treated as technical release. (See flow diagrams attached.)

- If changes are required to the technical content of the change after this event, a new change form will be required. (Do it right the first time - check point of no return.)
- The ECO number is placed on the change form after all Design Complete criteria is met — indicating to all involved that the change will be released and all affected can start their implementation efforts.
- If the change order requires revision the cognizant engineer and the CM technician will meet to determine if the correction was in the design or administrative data — thus determining who will write the ECO to correct the problem.

Procedure

- See the flow diagrams attached.

Primary Responsibility

- The CM function shall keep this standard current.

Authorization

____

Authorization VP Engineering

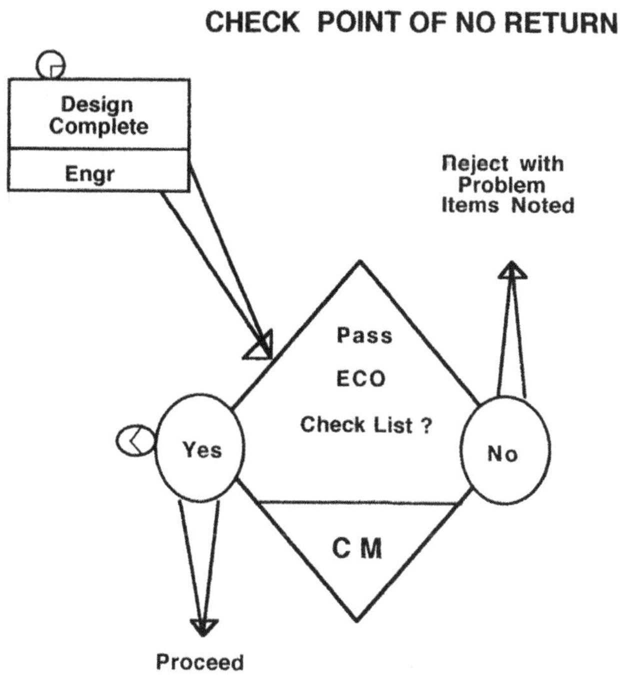
CHECK POINT OF NO RETURN
Design Complete
Engr
Reject with Problem Items Noted
Pass ECO Check List ?
Yes
No
C M
Proceed

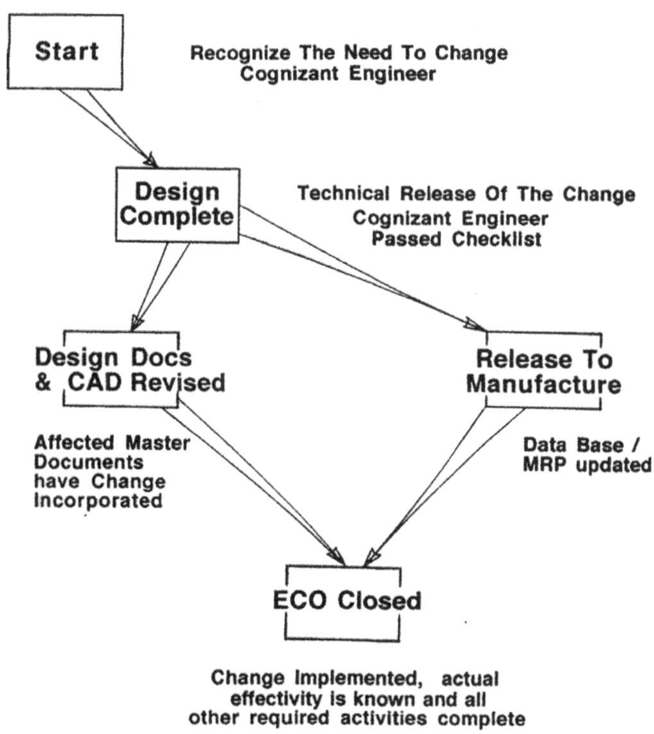
Start
Recognize The Need To Change
Cognizant Engineer
Design Complete
Technical Release Of The Change
Cognizant Engineer
Passed Checklist
Design Docs & CAD Revised
Release To Manufacture
Affected Master Documents have Change Incorporated
Data Base / MRP updated
ECO Closed
Change Implemented, actual effectivity is known and all other required activities complete

37

Actual Effectivity Tracking

Purpose

- To outline the method and responsibilities for tracking changes to their actual effectivity.

Applicability

- All changes to be processed by this CM function.

Policy/Practice

- Change processing is not complete when a change is released to manufacture — we must know the actual effectivity (traceability):
  1. For **non-interchangeable changes the exact units** containing the change (and thus those that don't) must be known.
  2. For **interchangeable changes the actual effective date** at a specified point in production is adequate — tracing to exact units is not necessary in most commercial environments. (Not necessarily true in some regulated environments - see FDA, DOD, NASA, etc., requirements.)
- **Examples of methods** for identifying the exact unit (non-interchangeable) are (pick one):
  3. Tracking to Serial Number is a very common method - "shop traveler" to Serial Number. When a change is incorporated, its number is recorded on the shop traveler. After the SN is assigned, the SNs with the change are recorded. Unit recycling may cause effectivity to be very mixed - Example: change #1326 effective at SNs 128, 130, 133, 136, 138 and up.
  4. "Make to Order" companies would typically make whole orders contain the change. If a change is to affect only part of an order, often a suffix digit is added to the order. If the ERP has an "Order Related BOM" capability, changes in the middle of an order could cause two Order Related BOMs to be created.

5. High end (high priced)/low volume manufacturers might accompany each unit with a BOM (and keep a copy on file), thus noting all BOM changes and non-BOM changes.
6. A new date code or lot number is assigned for each non-interchangeable change. A shop traveler – with date code/lot # attached to/accompanies each unit.
7. Units are tagged/marked, indicating a change is present ("Mod Identification").
   A01 – first production unit(s)
   A02 – first non-interchangeable change
   A03 – second non-interchangeable change
   B01 – first non-interchangeable change to be retrofit
   Recycling product issues solved by reassignment of Mod.
8. Each non-interchangeable change is assigned a Mod letter. A "scratch ticket" is attached to the product. When a change is incorporated into a unit the appropriate letter is scratched. A copy of each ticket is kept by SN as shipped. Mod scratch ticket:
   A B C D E F
   G H J K L M
   N O P Q R S
   T U V W X Y

9. Roll revision levels or part numbers and attach a traveler to each product to track revision level/part number. (Definitely not recommended, as Rev rolling and/or PN rolling are awful solutions. Can't tell a document only change from an interchangeable change from a non-interchangeable change. Sequencing issues are often resolved by revising revision levels. A traveler is still needed or all parts must be marked or tagged with a revision level.) Not recommended!

- Production Control shall be responsible for choosing the method of tracking with CM agreement.
- Production Control will be responsible for tracking all changes by the method chosen and entering the actual effectivity into the change form or notifying CM to do this.
- The identifier chosen (SN, date code, etc.) will be included on the product nameplate. If the scratch ticket is chosen — placing it near the nameplate is acceptable.
- If product and major assemblies are both serialized (often done to prevent "cannibalizing" for warrantee fraud), Production Control will be responsible for keeping track of actual SN(s) into SN and making that information available to the rest of the company.

Procedure

- See flow diagram, change form and change form instructions in other standards.

Primary Responsibility

- Production Control will be responsible for keeping this standard current.
- Production Control planners may take exception to this standard by noting on the change form why that was done.

Authorization

Authorization	VP Operations	VP Engineering

38

Line-Down Change

Purpose

- To describe the only acceptable method for emergency "line-down" changes — for multi-shift ops.

Applicability

- All emergency / line-down changes to be processed by this CM function. Note: Few companies need this process if they achieve a five day CM phase thru-put time with the ability to hand-carry in a half day.

Policy/Practice

- The cognizant engineer or an engineering representative shall be available to the production operations for every shift whenever the production line is scheduled to operate.
- A manufacturing engineer shall be available to the production operations whenever the production line is scheduled to operate.
- The CM process for hand-carry of changes in one half day must be in place and used when practical.
- The engineering representative and the manufacturing representative must agree on the need to process the change in less than one half day.

Procedure

1.	Production	Calls cognizant manufacturing engineer.
2.	ME	If a "line-down" condition exists, calls the cognizant design engineer.
3.	Cog Engr	Reviews conditions, if in agreement with urgency and the fix, makes two sets of marked-up documentation and signs both.

4.	ME	Signs both mark-ups.
5.	Cog Engr	Signs on to the system for change number and notes "line down" in the reason for change box. Posts change number to both sets of marked prints. Gives one set of marked prints to the ME.
6.	ME	Puts one set of marked prints on the production line.
7.	Prod	Incorporates change.
8.	Cog Engr	Completes change form online and delivers the other copy of marked prints to CM within one hour (second or third shift by 8 am the next morning).
9.	CM	If mark-ups and change form are not received by 8 am the morning after the number assignment, notify the VP of Engineering and VP of Operations. (They must personally let all engineers know that more than one hour is unacceptable.)
10.	CM	"Skips" all other signature boxes and incorporates the change (marked "line down") exactly according to marked prints. If there is a problem, another change will be required and that occurrence will be reported to the VPs..

Primary Responsibility

- The director of engineering services is responsible for keeping this standard current.

Authorization

Authorization	VP Engineering	VP operations

39
Closing a Change

Purpose

- To define the requirements for closing a change.

Applicability

- All changes processed by this CM function.

Policy/Practice

- Document Only changes will be closed after the Item Master and BOM have been updated and checked and the master design documents have been updated and checked — in either order.
- The Quality Function shall assure that the manuals match the product as shipped.
- Product changes will be closed only after all of the following tasks have been completed (modify to suit):
  1. The change has been implemented in the product as evidenced by the posting of the actual effectivity on the change form.
  2. The backup file of the updated master documents, parts lists and change form (as required) has been completed and stored off-site.
  3. A digital copy of the revised documents has been stored off-site.
  4. The field change package has been completed if applicable, modeled and signed as applicable.
- The CM function will then close the change form and enter the date of completion of whichever of the above requirements is completed last.

Procedure

- See the Change Form Standard and process flow.

Primary Responsibility

- The CM function shall keep this standard up to date.

Authorization

Authorization	Director of Engineering Services	Director of QA

40
Tracing Changes

Purpose

- To outline the methods and reports required in order to know the configuration of the product as it changes through time. In the DOD world, this is called "status accounting." It is necessary in order to do future troubleshooting, field change analysis, and minimize retrofit costs.

Applicability

- All changes to all products processed by this CM function.

Policy/Practice

- The configuration of the product will be available to all who need to know in three views: by change number, part number/parts list and by product.
- The "report" for the "by change number" view will be the change order form itself. The actual effectivity (SN – or whatever – on non-interchangeable and date on interchangeable changes) shall be entered into the change form for status reporting/troubleshooting purposes.
- The bill of material structure file is also a tool for tracing/status reporting for all parts list changes (interchangeable and non-interchangeable). The bill contains the current "as built" configuration and the "as planned" (as designed) configuration for any change wherein the future effective date feature is used.
- The revision block of an item's drawing or specification is a form of tracing/status accounting. The revision history must be kept for all changes. If the drawing is redrawn or the revision block truncated, the history must be kept in the "earlier" revision level document for tracking/troubleshooting purposes.
- The "description of change" in the revision block of all drawings and specifications must show the key words describing the change

(as opposed to "See ECO"). This will aid engineers in future troubleshooting efforts.

- The bill of material item master file will also contain all non-parts list changes by part number and revision level. Entered by CM.
- Other configuration status reports may be added as required by your kind of business and your regulators and for your warrantee purposes. The "report writing" feature of your system might be able to produce such reports.

Procedure

- None required.

Primary Responsibility

- The manager of CM shall maintain this standard in current condition.

Authorization

____

Authorization Engineering Services Director

Part VI

CHANGING PRODUCT SHIPPED - FIELD CHANGES

Definition:

Incorporating product changes into units already shipped.

41
Change of Field Units

Purpose

- To determine the impact of a change on the units which have shipped.

Applicability

- All changes which are processed by this CM group.

Policy/Practice

- If the units that have shipped are to be changed by retrofit or return goods repair, it shall not be a result of an interchangeable change.
- Field units to be retrofit should be a result of certain non-interchangeable changes. All non-interchangeable changes are not to be retrofit. Each must be considered on its own merits.
- Allowable categories of retrofit are defined as follows (determine your own categories):
  1. **Recall** – all units not containing this change must be found/returned to be retrofit. Typically limited to significant safety issues.
  2. **At Maintenance** – all units not containing this change must have it incorporated when the product is shut down for any regular or special maintenance.
  3. **Customer Option** – any unit the customer wishes to have this change in will be retrofit on a basis determined by the customer.
  4. **On Return** – to be retrofit on return to the factory for any reason.
- The field repair organization will retrofit on repair any or all categories of changes above as the customer wishes. The cost of retrofitting all units affected per the above will be included in the change cost calculations and estimates.
- No other changes will be retrofit on repair. Bringing all units to the "latest revision level" or installing changes not so designated is very expensive and will not be done.

- All failed items, worn items or damaged items will be replaced on repair to make the product "like new." The cost of such activity, however, will not be included in the change cost.

Procedure

- On any change checked "Retrofit: Yes," the cognizant design engineer must get the Service functions agreement and signature on the change form.
- The Service function will initiate the Field Change Form per that standard.

Primary Responsibility

- CM function will have responsibility for maintaining this standard.
- The cognizant design engineer and service representative may take exception to this standard providing they agree and indicate on the change form why that was done.

Authorization

____

Authorization	VP Engineering	VP Service

42
Field Change Policy

Purpose

- Define the company/division policy for affecting products with changes after shipment.

Applicability

- All products which have been production released by this CM function. Only those changes defined in Std #59 will be retrofit.

Policy/Practice

- A separate document (Field Change Order – FCO) shall be used to change the shipped units. (See Field Change Order Form Standard for the FCO format.)
- An FCO will be originated only because the design change order (ECO) so directs.
- Service and Design Engineering shall agree on the affect on the field (in the ECO) within the limits of these standards.
- Do not hold up the implementation of the design change order for the FCO writing and processing.
- Only non-interchangeable changes shall result in an FCO. Not all non-interchangeable changes will become an FCO. Each case will be considered on its own merits.
- Any significant safety issue will immediately become an FCO. Such changes will also cause an immediate letter to be sent to all customers as well as notification thru the mass media as required.
- The Field Support/Service function shall write the FCO for significant safety issues. The FCO shall be drafted for approval within 24 hours of the change order passing Design Complete and hand-carried through the process.
- A kit of parts (BOM) required to install the change in a finished unit shall be referenced in the FCO.
- A person other than the writer shall install one fix from the FCO and kit in a production unit. That person will mark up the kit and

FCO document. The kit list and FCO shall thus be "proofed/corrected" and distributed to the field.

- The FCO kit should contain a simple form (stamped, self-addressed postcard to CM) for feedback of serial number the field changed (unless you are "online" with the field).
- Don't close a design change order that affects the field until the FCO has been completed, modeled and approved by the writer and the "pilot installer."
- The FCO document shall be assigned a part number and included in the FCO kit (in order to allow for reorder in case of loss of the original).

Procedure

- See below for work flow diagram.

Primary Responsibility

- The director of engineering services shall keep this standard current.

Authorization

Authorization	VP Service	VP Engineering

FCO Flow Diagram

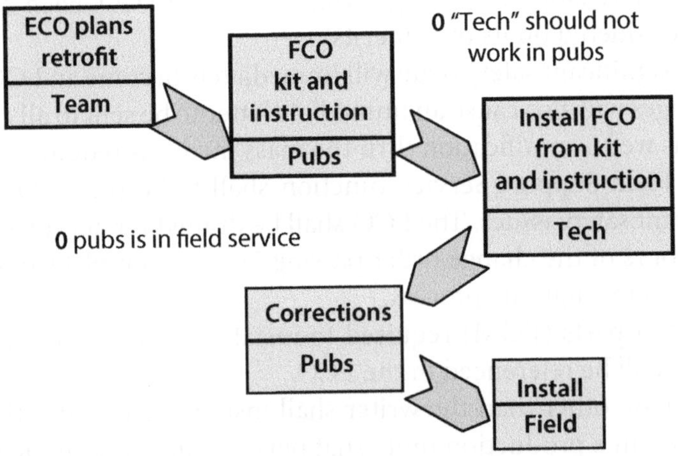

43
Field Change Order Form

Purpose

- To specify the only form to be used to change any products shipped.

Applicability

- All field changes resulting from design changes processed by this CM Function.

Policy/Practice

- The form on page 2 will be used for all field changes.
- Write your own form instruction. Place it on the back of hard copy forms and as a pop-up for the online form.

Procedure

- The Field Service Organization should develop any required flow diagram/procedure.

Primary Responsibility

- Service Organization will be responsible for keeping this form design current.
- Exceptions must be approved by the director of field engineering and the engineering services director.

Authorization

Authorization VP Service

Field Change Order **FCO#**

Choose One: Now ☐ On Failure ☐ At Regular Maintenance ☐

Product/Model Numbers Affected:

Manufacturing Effectivity: Ref ECO # Serial Numbers without (needing) change:

FCO Kit Part Number: **Estimated Time to Install:**

Expand on the Following Subjects in Order:

1. Reason/Justification for Change.
2. Publications Affected.
3. Description of Change.
4. Change installation step-by-step procedure.
5. Special tools required (must be available at site or in the kit).
6. Removed parts and field warehouse parts disposition.
7. Test/check-out procedure.
8. List of attached and referenced documents.

Authored By: **Accepted By:**

Date: Page of Pages FCO Part Number Rev

Form #

Part VII
ODDS AND ENDS

Definition:

Those standards not fitting neatly in any other chapter.

44

Costing Design Changes

Purpose

- To outline the changes that must have their total cost calculated, the responsibility for calculating the cost, to indicate those costs that must be included, specify the form used, furnish form instructions and additional change approvals required based upon the cost.

Applicability

- Certain design changes to be processed by this CM function.
- Changes which must have their total cost calculated prior to approval or release are those made to "reduce costs" and to "improve performance" and others as directed by the Request Review Team or the Change Team.
- Changes that are totally or partly "to meet specification" need not have the cost calculated unless the team requests it or unless the cognizant design engineer has alternative designs to cost.
- Document Only changes shall not be cost calculated.

Policy/Practice

- The team can decide not to approve a change until the cost has been calculated. The team may also decide to approve a change subject to completion of the cost calculation / payback prior to release.
- Cost calculating will be done by CM (Industrial Engineering/Cost Accounting are alternates - see the EDC Handbook for a discussion of alternatives).
- All the cost directly associated with a given change shall be included in the cost — by Activity Based Cost (ABC). Costs/savings to be included (but not necessarily limited to):

Unit Costs:

- Old part/new part standard material and labor cost/savings - at IE/Accounting standard cost.
- Assembly/test labor cost/savings - at IE/Accounting standard cost.

- Production material/labor burden (overhead) ABC factors applied.
- Unit costs above will be annualized by the expected annual production quantity of the product(s) affected.

Startup Costs:

- Supplier rework, tooling, premiums and cancellation charges.
- Manufacturing scrap, rework, tooling, fixtures, test equipment, documentation and inspection equipment costs.
- Field Change Order kits, service labor, returns repair labor, spares crib scrap/rework.
- Design Engineering, CM, Design/Drafting labor.
- Engineering labor burden (overhead) ABC factors applied.

- Changes called **"cost reductions" must pay back one-time costs within ___ months** or they will be canceled/rejected. (See the EDC Handbook for guidance.)
- If "improvements over/above specifications" add more than $ ___ per unit (applicable to the products listed in Product Improvements Standard) they must be approved by the chief engineer.
- Product improvements must have management approval based on the one-time startup costs as follows (pick you own $):
  1. Up to $5,000 may be approved by the first level Engineering Manager.
  2. From $5,001 to $10,000 may be approved by a second level Engineering Manager.
  3. Changes exceeding $10,000 must be approved by the VP of Engineering.
- The Production Control Planner will furnish the estimated planned effectivity and part disposition to CM for cost basis (in this sense, this will be an estimate, not a calculation.)

Procedure

- See work flow diagram on page 3.
- Accounting must furnish up-to-date ABC factors.
- Cost Calculation Form and Form Instructions are on page 4 and 5 of this standard.

Primary Responsibility

- The CM function will have responsibility for keeping this standard current.

- Exceptions to this standard may be taken by the change request or change team by stating in the applicable change order form why it was necessary.

Authorization

Authorization	VP Engineering	VP Operations

Cost Flow Diagram

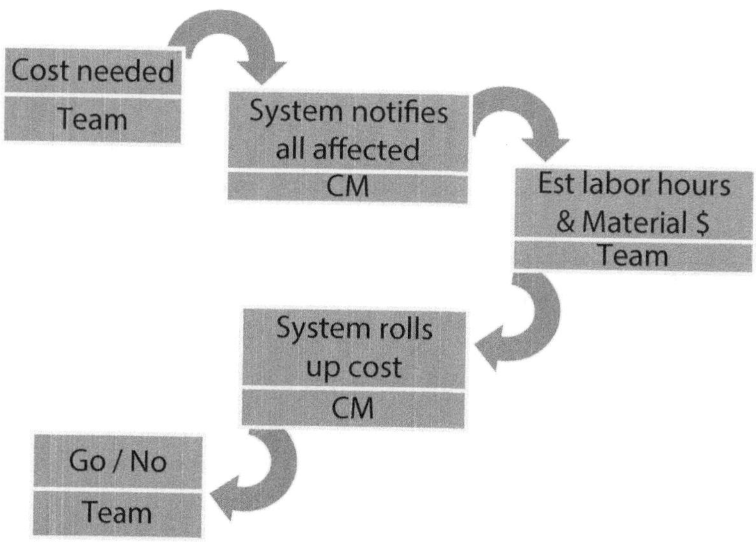

Request / Change Cost Estimate **EC3 Corp** © Pg 1 of 2

General:

☐ Request ECR# ______ ☐ Change Order ECO# ______ ☐ Other #____

Change type / class ______________________________

Key Estimator Name ______________________________ Date __-__-____

Title of Request / Change ______________________________

Initial / Rev Date __-__-____ Reason for Revision ______________________________

Unit Part Cost

Old Part:

Part Number	Std $	Qty Per Unit	$ Per Unit	Scrap Qty	Scrap $	Rework $ Per Part	Qty to Rework	$ of Rework
____________	____	____	____	____	______	____	____	_____
____________	____	____	____	____	______	____	____	_____
____________	____	____	____	____	______	____	____	_____

V **Old Part $ / Unit** ______

Total Scrap $______ Total Rework $ ______

New Part:

Part Number	Std $	Qty Per Unit	$ Per Unit	Qty in Field Kit	Field Kit $	Supplier Tool $	Company Tool $
____________	____	____	____	____	_____	______	______
____________	____	____	____	____	_____	______	______
____________	____	____	____	____	_____	______	______

V **New Part $ / Unit** ______

Total Field Kit $ ______

Total Supplier Tool $______

Total Company Tool $ ______

Difference between Old and New Part Cost x Mat'l Overhead ___% = **Material $ Diff / Unit** _______
(parenthesis if old is less than new)

Unit Labor

Direct Assembly Labor Difference:

_____ Hrs x Labor Rate ___.__ $/Hr x Overhead ___% = Assembly $ Diff / Unit _______
(if old is less than new)

Direct Test Labor Difference:

_____ Hrs x Labor Rate ___.__ $/Hr x Overhead ___% = Test $ Diff / Unit _______
(if old is less than new)

Other ____________________ **Labor:**

_____ Hrs x Labor Rate ___.__ $/Hr x Overhead ___% = Other $ Diff / Unit _______
V (if old is less than new)

Request / Change Cost Estimate **EC3 Corp ©** Pg 2 of 2

Annualizing Unit Costs

Product Per Unit Savings / (Cost Increase): from above = $______

Annual Schedule Quantity = ______

Annual Product Unit Savings / (Cost Increase): = $________

Implementation ("one time" or "start up") Costs

Materials:

Scrap $______ + Rework $______ + Other $ ______ = $_______

Suppliers Charges:

Rework $ Scrap $ $ Tool Cancel/ Prem $ Returns Cr $

_____ + _____ +_______ + _____ - _____ = $_______

Manufacturing:

Tool $ Test Equip $ Facility $ Other $

______ + ______ + ______ + ______ = $_______

Engineering:

Engr _____ Hrs x $ / Hr _____ = $________

CAD Design / Drafting:

Des Dftg _____ Hrs x $ / Hr _____ = $________

V

Total Design & Development = $_______

Field / Repair:

Kit $ Labor _____ Hrs x $/Hr _____ Rewk $ Scrap $ Returns Credit

______ + _____ + _____ +_____ - _____ = $_______

Configuration Management:

CM ______ Hrs x $ / Hr _____ = $ _______

Quality Assurance:

QA ____ Hrs x $ / Hr _____ = $ _______

V

Total Implementation Costs = $________

Pay Back / Opportunity Cost Analysis **(Company pay Back Policy = __.__ Years)**

This Request / Change Pay Back in Years (cost increase in parenthesis) = $__.__

Opportunity Cost (what other project might limited resources be better invested in): ________________

Conclusion:

Cost analysis indicates: ☐ **do change** ☐ **don't do change** ☐ **Other**

Estimators Comment: ____

____

45
Data Dictionary

Purpose

- To define each element of data in the BOM and in Design Documents, their source and the function responsible for entry into the database.
- To use this technique for all programs relating to Design Data. To avoid multiple definitions for the same data element. One government site reported 27 different definitions for a part number.

Applicability

- All data to be entered into the Engineering/Manufacturing databases.

Policy/Practice

- Determine whether an element of data relates to a document or an item.
- Define basic information as in the following example:
  Data Element: Item weight
  Source: Released part document – furnished by the cognizant engineer
  Entered by: CM technician
  Character Definition: 5 digits – NNN.N (N = Numeric)
  English definition: The weight of the part in pounds and tenths of pounds.
  Notes: Not to be entered for assemblies. Not to be entered for documents unless they are shipped with the product. Allows "roll-up" of assembly or product weight even before the product is built.
- This element is shown, for example, not to encourage its use, as the product can more easily be weighed after build.

Procedure

- None required.

Primary Responsibility

- The CM Manager will be responsible for keeping this standard current and must obtain the IT Department co-operation.

Authorization

Authorization VP Engineering

46

System Measurements

Purpose

- To identify the major events in the Release, Request and Change Processes to be measured and the minimum reports to be made.

Applicability

- All Releases, Requests and Changes processed by this Document Control function shall be measured. See the author's CM Metrics for many other specific measurements.

Policy/Practice

- At a minimum, the process speed, volume and accuracy shall be measured and reported to the executive management as well as other involved management and key people.
- Measurements will be published at least monthly, expressing speed in graphical form in work days or work hours (lapsed time).
- The process work flow diagrams found in each process standard show the minimum events to measure process time and volume by a "clock" attached to the event.
- Those points will be recorded in the related forms with the date the event is completed. If "finer than calendar days" are to be measured, the forms must include the time of day the event was completed.
- A method of measuring the accuracy of each process will be developed. For example: the change process will count revisions to the change — both technical and administrative (attached report). The revisions to the change will be expressed as a percent of the total changes during the same period of time.

Procedure

- Measure the process time between each "clock" in every work flow and report at least monthly. (See attached example report.)

- Measure the volume “passing” each clocked event and report at least monthly. (See attached example report.)
- Report on the accuracy/quality of each work flow and report at least monthly.

Primary Responsibility

- The CM manager is responsible for this reporting.

Authorization

Authorization	CM Manager

Example of Change Process – CM Phase – Time and Volume Report

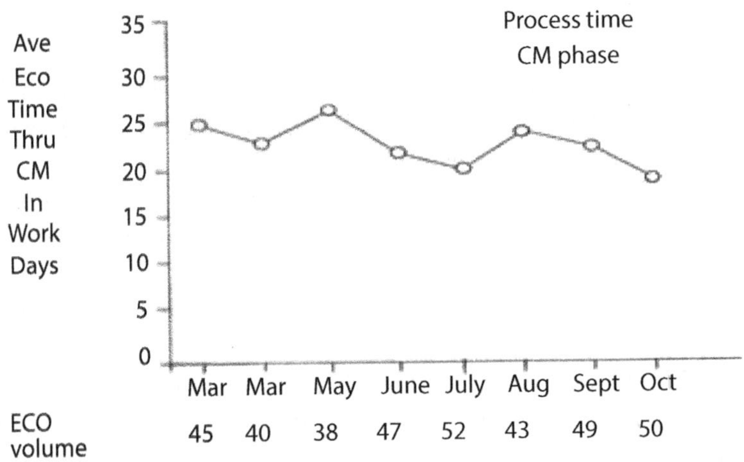

Example of Change Process Quality Measurement (changes to the change for technical reasons) (See CM Metrics)

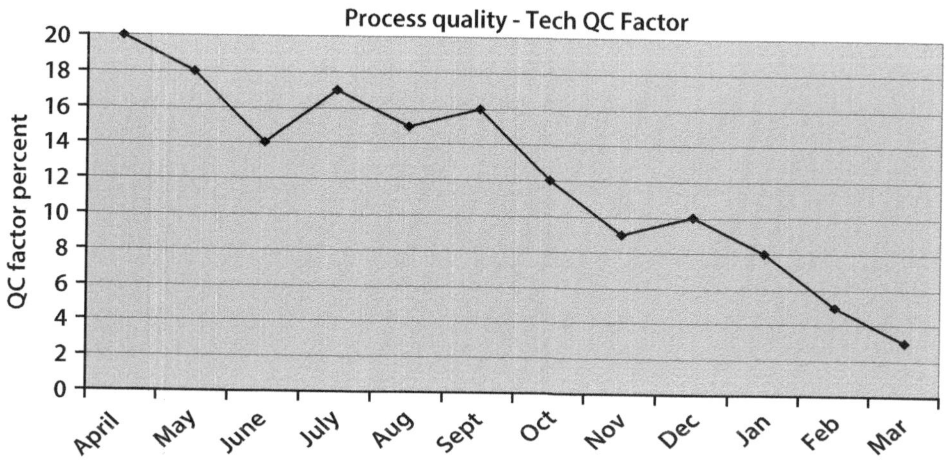

47

Users Guide/CM Plan

Purpose

- To summarize the standard practices to be used in Engineering Documentation Control (EDC) / Configuration Management (CM), for training company people and to give to auditors, agencies, customers and suppliers.
- These standard practices cover the four major CM processes: Document & Release, BOM, Request, and Change.

Applicability

- All product design documentation processed by or affecting this CM function.

Introduction

- This system can and should be edited to suit your company environment, organization and terminology.
- This system is totally integrated, comprehensive, separated into unique standards for each single subject and has many notes to encourage editing for your unique operations.
- These concepts have helped companies to drastically reduce their process time and errors. Three workdays to release and five to six workdays average to change (with the ability to hand-carry in one half day) through the Doc Control part of the process.
- One person is designated as the CM manager. That person would, among other duties, oversee the editing of the Standards Manual in order to assure best currently attainable processes.
- These standards should be compatible with any ERP, PDM, CAD, etc., system but care should be taken to edit to fit with those systems' use and terminology.
- This summary document should be created after implementation of those standards which can be readily implemented with the current situation at your company. It should be updated after every continuous improvement project.

- Do not expect any of the standards to be "static" but rather a continuously changing set while your processes are continuously improved. Update this document accordingly.

Concepts

- Basic elements must be in place as building blocks for the release, BOM, Request and Change Control processes. Those standards are listed in the Contents pages.
- Use of cross-functional teams in all the EDC processes is encouraged in order to achieve "buy-in" from key functions and to achieve manufacturability, testability, etc. The standards for such team activity are:

05	Teams for All CM Processes
19	Team in Release
22	Team in Request
25	Team in Change

- Team meetings in the release process and especially early in the design phase are used to discuss problems, manufacturability suggestions and cost reductions just as soon as they are identified. The Technical signatures are obtained by the cognizant engineer at the team meetings on the documents themselves. The traditional "Change Control Board" has been eliminated in favor of a technical review meeting and a carefully specified change team and process.
- There is a careful separation of Technical Documents into categories: Product Design, Service Support, Manufacturing Process, Sales, Quality and EDC Process documentation. We do not "bundle" with the design documentation in the processes. This separation allows:
  - Release of manufacturable Design Documentation in lead time to build or buy and without holding them up for other documentation changes.
  - Design document changes are fully defined during the Change Design Phase and incorporated into the masters during the CM phase.
  - Service Support technical documentation is separately controlled by the Publications function, Manufacturing Process technical documentation by the ME or IE function, Sales technical documentation by the Sales Support function and Quality technical documentation by the QA Engineering function.
  - Support, Manufacturing Process, Sales or Quality document changes (if affected by a design document change) are planned

during the Change Design Phase, started during the CM Phase, and completed during the Implementation Phase.

- Support, Manufacturing Process, Sales or Quality document changes, no matter the source, are not done by the same process / form(s) as the design documentation. Other process(s) and form(s) have been designed for those document changes. Such standards are approved by the CM manager and placed in this manual.
- CM/EDC process documentation (these standards) are maintained online in a secured manner by use of a separate process, without use of the Release or Change form or processes.

- We have online work flow and forms so the action documents (Release, Request and Change forms) are distributed online, not in hard copy — the first step toward "less paper." The released or change documents themselves are also not distributed in hard copy. The people and functions that require copies of the released or changed documents will obtain those documents online (via CAD, PLM or ERP) or in hard copy when and as needed ("pull system" not a "push system").
- This company has three major phases of new product development. They are called Design & Development, Pilot and Production. Design Engineering has more detailed phases for new product development, but from a company perspective these three phases (coded E,S,P) will be identified on the design documents (rev date, numeric and alpha) and in the ERP and PLM system. Thus, one can tell the phase that an item can properly be valid for by looking either at the document or the system.
- The same form is used to release new designs and to change existing designs. The form instructions address both. The standards that support the Release process:

18	Release Policy
19	Teams in Release
20	Phase Release

- The Bill of Material process (not exactly a process) is addressed in conjunction with the ERP/PLM systems. Standards clarifying the Engineering-Manufacturing interface and the requirement to approach a single BOM structure are:

08	Quantity and Unit of Measure
09	Bills of Material
10	Approved Manufacturers List

- Any employee can request a design change/improvement/cost reduction. Requests are not changes and will not necessarily become changes. The request form and work flow process are in the following standards:

21	Change Request Policy
22	Team in Request
23	Request Form

- The cognizant engineer has change control authority during the Design & Development phase. That engineer is the only person that CM/EDC will give a request to or accept a change from. A list of cognizant engineers is maintained in CM.
- All changes to Pilot and Production released items are only done by the process described in the change standards. The standards that support the change process are:

24	Change Control Policy
25	Team in Change
26	Change Form
27	Interchangeability
28	Part Number & Revision Level Changes
29	Change Classification
30	Mark Up of Design Documents
31	Effectivity
32	Effectivity Management
33	Disposition of Old Design Parts
34	Impacted/Affected by a Change
36	Change Design Complete
37	Actual TrackingEffectivity Tracking
38	Line-Down Change
40	Tracing Changes

- The change process and responsibilities are divided into three major phases. The CM manager has overall responsibility for the process design for all phases. The basic responsibility for the fast and accurate processing of the Release or Change is as follows:

  1 Change Design Phase = Engineering Responsibility
  2 Doc Control / CM Phase = CM Responsibility
  3 Implementation Phase = Manufacturing Responsibility

- During the pilot phase the normal signature requirements can be shortened to only the cognizant engineer and manufacturing engineer — all other signatures can be skipped, but are informed and have "exception ability."
- Engineering Change Classification is uniquely defined on the change form.

- Carefully specified use of marked-up design documents in a change is encouraged in order to prevent transposition errors and wasted time.
- Certain functions can originate a change for the cognizant engineer, using the change form, if a solution to the problem is known and agreed upon. This avoids many change requests and shortens the total process time.
- Only the change form can be used to change a design and associated documents when a document is in pilot or production phases. Change Request, Deviations, and other forms or processes are not used to change a document, item or product.
- Effectivity of the change by product affected is planned, re-planned as needed and the actual effectivity captured. This planning is done by (date, order, serial number, etc.). A unique point in the production process (pipeline) may also be specified. Production Control manages the effectivity plan, re-plan and tracing to actual.
- There is a "point of no return" in the process called "Design Complete." There is a checklist requirement at this point according to the Change Design Complete Standard. This prevents the engineer from prematurely entering changes into the CM process which have not been modeled, tested, technically reviewed, complete, etc. After this point the change will proceed as designed and will not be held up, technically revised, expanded or other changes "piggybacked" on it. If the change is not technically correct another change form will generally be required to correct the problem.
- Specific provisions are made for disposing of old design material affected by each change. This prevents the accumulation of such material in "bone piles" in manufacturing.
- Closing a change form is done only after specific actions have been completed. The standard that supports this concept is:

  39 Closing a Change
- Configuration status accounting / reporting is accomplished by:

  40 Tracing Changes
- Field Changes are limited and addressed as an integral part of the change process. The change form must determine whether or not the change will be retrofit and on what basis. Before the change is closed, the field change form must be released. Standards which support field retrofit are:

  41 Change of Field Units

  42 Field Change Policy

  43 Field Change Order Form

- Costing of certain changes (ABC – Activity Based Cost) is provided for. The standard that describes that activity is:
  44 Costing Design Changes
- In order to minimize thru-put time, the work flow diagrams have expected average time entered on the diagrams or in the standards.
- A Configuration Management Technician monitors the entire process and aids the engineers thru the design release / redesign processes.
- When customer or regulator notification or approval is required the process starts early and has provisions for proceeding with measured risk. **Note: Not specified in the process, but done at many companies, is a contractual clause to limit the time allowed for customer approval on non-interchangeable changes. (If approval / disapproval is not received in ten workdays the change is approved by default.)**
- Terms and acronyms are precisely defined in standard #48. These terms and their use in every standard have been carefully edited.
- The process flow time, volume and quality are measured and reported to management, involved parties and interested parties. CM distributes those reports on a monthly or weekly basis. Speed of the processes is critical to profitability.
- Some other standards are included to enhance the basics and system processes.

Procedure

- See the process flow diagrams and forms attached to or in the applicable standard.

Primary Responsibility

- The CM manager is responsible for keeping this standard up to date.

Authorization

Authorization VP Engineering Date

48
Acronyms and Definitions

Purpose

- To define the terms and acronyms used in this Standards Manual.

Applicability

- This Standard applies to all CM policies, procedures, SOP, flow diagrams, forms and form instructions.

Policy/Practice

- Sources are stated, or are the author's definitions, or the author's modification of other's definitions.

Assembly
Any make item with a parts list or purchased item with a supplier parts list.

Benchmark
A fixed point of reference or a standard for comparison. An outstanding example that is appropriate for use as a model.

Block Diagram
An illustration in which essential units of any system are drawn in the form of blocks and their relationship(s) to each other are indicated by appropriate interconnecting lines. Appropriate for dog and pony shows. Not a work flow diagram.

BOM – Bill of Material
A compilation of the parts lists for an assembly or a product. May be displayed in a variety of formats.

DoD terms:

As Designed BOM	A DoD term that means different things to different people.
As Planned BOM	A DoD term that means different things to different people.

As Built BOM A DoD term that means different things to different people.

As Shipped BOM A DoD term that means different things to different people.

As Maintained BOM A DoD term that means different things to different people.

See DoD specs for their stated meanings. Realize that one might be talking about a BOM for each unit of product or a BOM for each order of product, etc. Also keep in mind that ERP systems generally reflect effectivity dates (planned or actual) as that date wherein the old design part will no longer be issued from stock and the new designed part will. Such a date may have little resemblance to the date of build or shipment.

CAD – Computer Aided Design / Drafting
Computer tools used to assist in the creation and maintenance of configuration – design definition and documentation. Source: EIT Glossary

CAE – Computer Aided Engineering
Computer tools used to design and conduct analytical tests on design models. Source: EIT Glossary

CAM – Computer Aided Manufacturing
Computer tools (which may be capable of extracting and utilizing the CAD data) used to assist in the creation and maintenance of the methods used to produce the item. Source: EIT Glossary

CCB – Change Control Board or Configuration Control Board – A team of people too often involved too late in the change process.

CDR – Critical Design Review

CDRL – Contract Data Requirements List

CIM – Computer Integrated Manufacturing

Class I
Categorization of non-interchangeable changes and other changes as defined by the DoD.

Class II
Categorization of product changes that are not class I.

CM – Configuration Management
A simple, fast, accurate, systematic, efficient, measured and well-understood process approach and discipline to plan, identify, control and track a products configuration throughout its life. Source: Frank Watts

CNC – Computer Numerical Control

Concurrent Engineering
A management / operational approach which improves product design, production, operation and maintenance by developing environments in which personnel from all disciplines (design, marketing, production engineering, process planning, and support) work together and share data throughout all phases of the product life cycle. Source: EIT Glossary

Configurator
Software tools that simplify order entry by asking which features and options the customer wants, then they may apply predefined rules to correctly configure the product with the desired attributes, &/or test for rules conflict, &/or launch the order and may generate the BOM, routing and price. Source; Dick Bourke and Associates Consulting

Data Element
The smallest pieces of data that cannot be subdivided and still retain any meaning. The terms "data item" and "data field" or simply "field" are often used synonymously with data element.

Database
A collection of structured data elements which are application-independent.

DBMS – Database Management System

Deviation
Defines the conditions and parameters for a variance from the item / product design or process specifications, normally for a specific time frame or specific number of units. Documented with a form. Should be posted in the document revision block.

DFT – Demand Flow Technology – also see JIT

Dispositioning (CM Slang)
Determining what to do with the old design parts resulting from a design change.

Doc/Document

1. Sometimes used as a reference to a change that affects only the document (Doc or Records Change).

2. Usually refers to technical, product-related documents. Includes, but is not restricted to design drawings and specifications. It is not media specific, and some documents may never take the "traditional form" of "formatted paper."

Document Representation
A set of digital files which, when viewed or printed together, collectively represent the entire document. (For example, a set of raster files or a set of IGES files.) Note: A document may have more than one document representation. Source: Mil-Std-2549

Document Management System
Allows users to store, search, and manipulate documents electronically and to maintain a library of text and images in a compact space. Many such systems also provide a means for passing documents across a network.

DoD - Department of Defense

DoE - Department of Energy

ECO - Engineering Change Order
Term used to refer to any form utilized to specifically define the change and to accomplish the evaluation, review, approval, planning and incorporation of the change.

ECN - Engineering Change Notice

1. Another generic term for an ECO.
2. DoD term used when the organization writing the ECO does not have the master in its facility to incorporate the change and sends a version of the ECO (the ECN) to the owning facility to direct master update.

ECR - Engineering Change Request
Term used to refer to any form used to solicit the responsible engineer to take ownership of a problem or suggestion for possible change.

ECP - Engineering Change Proposal
Term used to refer to any form used to obtain customer review and/or approval of a change.

EDC - Engineering Document Control
A discipline similar to CM but is focused on an organization's documents

Effectivity (CM Slang)
A method of specifying when (usually a date) and at what point in the manufacturing process a design change will be made and/or which units have the change installed (SNs for example).

ERP – Enterprise Resources Planning
Software programs that are an extension of the Manufacturing Resources Planning (MRP) concepts. These systems are to take, make, ship and account for customer orders and thus to automate and integrate materials, business, accounting and production management processes.

FCA – Functional Configuration Audit

FDA – Food and Drug Administration

FCO – Field Change Order
A method or form for incorporation of a design change in units that have been shipped.

Firmware
Computer programs, instructions or functions implemented in user-modifiable hardware. Such programs or instructions — stored permanently in programmable, read-only memories — constitute a fundamental part of a hardware product.

Flow Manufacturing
Other names: Agile manufacturing; Continuous (flow) manufacturing; Demand-based conversion; JIT (just-in-time) manufacturing; Lean manufacturing;

- A production technique in which the work moves from one operation to the next continuously (one piece at a time).
- Usually requires the rearrangement of the plant into lines and/or cells.
- Simplified scheduling and reporting appropriate for short production lead time.
- Demand-based flow links production schedules to customer demand for quick-response and short lead time.
- Many of the advantages of mass production without the long production runs; enabler for mass customization.
- Often incorporates just-in-time and Kanban techniques for material control.

Flow Chart
Diagrammatic representation of the operations involved in a process or system. Flow lines and arrows indicate the sequence of operations and/or the flow of data or forms.

Flow Diagram
Same as a flowchart with the added feature that the party/department which performs the operation is designated in the diagram. Inclusion of the responsible function in the diagram is what distinguishes them from Block Diagrams or Flowcharts.

GMP – Good Manufacturing Practices

GT – Group Technology
The grouping of discrete items into families having similar design and/or manufacturing process characteristics. Source: EIT Glossary – See also CSM

I – Interchangeable
See Standard #43

Interface
A shared boundary between two pieces of equipment or computer software code.

The hardware and software needed to enable one device to communicate with another.

IE – Industrial Engineer

Imaging
The electronic representation and storage of documents. Imaging systems typically allow retrieval and dissemination of those documents. Such systems may be thought of as electronic vaults.

Item
A generic term for any part, component, assembly or product. Might also be used to refer to documents. Generally limited to things assigned part numbers.

JIT
Just-in-Time is an approach to manufacturing that stresses the benefits inherent in a "Pull" system, wherein a small amount of material is brought to the work and placed in a Kanban only when it is needed. To achieve this goal each operation must be synchronized with those before and after it. Also called short-cycle manufacturing, demand flow technology, stockless production, zero inventories, etc. One primary goal being the reduction of inventory and inventory carrying costs.

JPEG – Joint Photographic Experts Group
A bitmap (raster) graphics format.

KISS – Keep It Swift and Simple or Keep It Simple Stupid

Kanban
Japanese terminology for one element for achieving JIT – the term specifically relating to the use of a location / card to indicate part status/quantity needs.

MC – Material Control

MES – Manufacturing Execution Systems
Software to track work-in-process through detailed product routing and tracking, labor reporting, resource and rework management, production measurement and data collection. These systems are similar to ERP but generally used in process manufacturing (food, drug, beverage, etc.).

MRP
Software programs that perform a variety of functions to aid in the manufacturing process. They use the Bill of Material, inventory data and the master production schedule to calculate requirements for materials and orders (MRP).

Software programs which do the above and include planning use of company resources, suppliers, equipment and processes, and sometimes financial and distribution management.

Improvements to MRP (MRP II). Subsequent improvements are termed ERP.

NASA – National Aeronautics and Space Agency

Network

1. Any system of computers and peripherals.
2. Any combination of circuit elements in an electrical or hydraulic circuit.

NI – Non-Interchangeable
See Standard #43

NRA – Nuclear Regulatory Agency

Parent/Child
Descriptive term that represents the assembly-component relationship in the parts list or BOM.

Part
Any physical item without a parts list. Might also be called a piece part, material, item or component. A part or component to us could be an assembly or product to our supplier.

Part Class or Part Classification
Methods to classify parts and documents of a product by their function or by the processes used to manufacture them. (See also Group Technology)

Part Master/Item Master
Data/information related to the part or item regardless of where it is used.

PC – Personal Computer or Production Control

PCB – Printed Circuit Board

PDF – Portable Document Format
File format for distributing formatted documents, with fonts & graphics, in a read & print form that can be read by Adobe Systems, Inc., reader. The user does not have to have the originating application.

PDR – Preliminary Design Review

PL – Parts List
A list of parts, sub-assemblies and usually reference documents for an assembly.

PLM – Product Lifecycle Management
Software that helps to manage design documents and data. Might also be referred to as Product Data Management (PDM), Product Information Management (PIM), Engineering Data Management (EDM), Engineering Information Management (EIM), or generically as File Control or Vault.

PN – Part Number

Process Map – See block diagram.

QA – Quality Assurance

QC – Quality Control

QE – Quality Engineering

Records Change
A term used to refer to a change, which affects only the document(s) and has no affect on the physical product. Sometimes referred to as a "Doc" Change or "Doc Only" Change.

Release
The act of obtaining the required approvals to progress from one phase of design or manufacturing to the next phase.

Revision
A modification of any product technical data or document. See Standard #44.

SCO – Software Change Order

SN – Serial Number

TE – Test Engineer

TIF – Tagged Image File Format (TIFF)
A bitmap (raster) graphics format.

Vault
Print/document storage, either electronic or physical. The PLM system's computerized data storage area and databases. Information stored in PLM vaults is controlled by a system of rules and processes.

Version
Non-interchangeable changes to a hardware or software item. Generic term referring to iterations of a document or item.

WIP – Work in Process

Workflow
Any diagram depicting the flow of work. The software ability to graphically designate and change the distribution, routing for adding value and/or approval routing of documents related to a business process.

Procedure

- None required.

Primary Responsibility

- The CM Manager will keep this standard up to date.

Authorization

____

Director of Engineering Services

Author Biogaphy

Frank Watts BSME, CCDM

Frank is a product of a University of Illinois engineering education and has over forty-eight years of industrial and consultation experience. He founded his own company to specialize in configuration management with specific expertise in the basics of product release, change control, product lifecycle management, bills of material and other Engineering Documentation Control, Configuration Management and Product Data Management issues.

Formally a Director of Engineering Services, Director of Operations and a Director of Manufacturing Engineering, Frank has also managed Configuration Management, Cost Estimating, Industrial Engineering, Test Engineering, Pilot Production and Repair Operations. He has contributed as a Design Engineer, Industrial Engineer, Manufacturing Engineer, Systems Analyst and Project Manager.

Frank has guided the development of the engineering change control process at many companies and made significant contributions toward improving the new product release process, installed ERP systems, new numbering systems, helped companies attain a single BOM database and guided reengineering of the CM processes.

His industrial experience was with Caterpillar, Collins Radio, Control Data, Storage Technology, UFE and Archive. He has presented seminars on the subject to over 4,000 product manufacturing people. He is an NDIA Certified Configuration and Data Manager, author of several magazine articles and author of ***Engineering Documentation Control Handbook*** (Elsevier). This book (now in the forth edition) has sold over 13,000 copies — quit counting because the publisher doesn't count e-books. According to one reviewer, it "Sets the Stage for Innovation." Frank has also authored ***CM Metrics*** (Elsevier) first edition published in 2009. Most recently Frank authored ***CM for Senior Managers***.

Credits:

- Over 75 clients from all segments of product manufacturing
- BSME/IE – University of Illinois
- Graduate Studies – University of Iowa
- Over forty professional courses and seminars.
- Articles – Production and Inventory Management Journal, Midrange ERP, and the ACDM Journal. 48 Ways to Faster Design

Changes, copyright 1994, second edition, 1998. 54 Trolls Under the CM Bridge, copyright 2000. Many white papers.
- Benchmarking surveys – in the EDC Handbook.
- Past member – ASME, SAE, AIIE, ADPA, APICS, ACDM.

He welcomes email and phone calls. Will offer free advice (which may be worth just that) for brief questions.

P O Box 205	**970-887-9460**	**wattsec3corp@gmail.com**
		Winter Park CO 80482

Printed and bound by CPI Group (UK) Ltd, Croydon, CR0 4YY
17/09/2026
14964250-0002